Field Testing Genetic
Organisms:
Framework for Decisions

Committee on
Scientific Evaluation of the
Introduction of Genetically Modified
Microoganisms and Plants
into the Environment

Board on Biology
Commission on Life Sciences
National Research Council

NATIONAL ACADEMY PRESS
Washington, D.C. 1989

National Academy Press • 2101 Constitution Avenue, N.W. • Washington, D.C. 20418

NOTICE: The project that is the subject of this report was approved by the Governing Board of the National Research Council, whose members are drawn from the councils of the National Academy of Sciences, the National Academy of Engineering, and the Institute of Medicine. The members of the committee responsible for the report were chosen for their special competences and with regard for appropriate balance.

This report has been reviewed by a group other than the authors according to procedures approved by a Report Review Committee consisting of members of the National Academy of Sciences, the National Academy of Engineering, and the Institute of Medicine.

The National Academy of Sciences is a private, nonprofit, self-perpetuating society of distinguished scholars engaged in scientific and engineering research, dedicated to the furtherance of science and technology and to their use for the general welfare. Upon the authority of the charter granted to it by the Congress in 1863, the Academy has a mandate that requires it to advise the federal government on scientific and technical matters. Dr. Frank Press is president of the National Academy of Sciences.

The National Academy of Engineering was established in 1964, under the charter of the National Academy of Sciences, as a parallel organization of outstanding engineers. It is autonomous in its administration and in the selection of its members, sharing with the National Academy of Sciences the responsibility for advising the federal government. The National Academy of Engineering also sponsors engineering programs aimed at meeting national needs, encourages education and research, and recognizes the superior achievements of engineers. Dr. Robert M. White is president of the National Academy of Engineering.

The Institute of Medicine was established in 1970 by the National Academy of Sciences to secure the services of eminent members of appropriate professions in the examination of policy matters pertaining to the health of the public. The Institute acts under the responsibility given to the National Academy of Sciences by its congressional charter to be an adviser to the federal government and, upon its own initiative, to identify issues of medical care, research, and education. Dr. Samuel O. Thier is president of the Institute of Medicine.

The National Research Council was organized by the National Academy of Sciences in 1916 to associate the broad community of science and technology with the Academy's purposes of furthering knowledge and of advising the federal government. Functioning in accordance with general policies determined by the Academy, the Council has become the principal operating agency of both the National Academy of Sciences and the National Academy of Engineering in providing services to the government, the public, and the scientific and engineering communities. The Council is administered jointly by both Academies and the Institute of Medicine. Dr. Frank Press and Dr. Robert M. White are chairman and vice chairman, respectively, of the National Research Council.

This study by the Board on Biology was funded by the Biotechnology Science Coordinating Committee composed of the Department of Agriculture, Environmental Protection Agency, Food and Drug Administration, National Institutes of Health, and the National Science Foundation, under grants numbers BBS 8820985 and 12-34-30-0024-GR.

Library of Congress Catalog Card Number 89-63061

International Standard Book Number 0-309-04076-0

Printed in the United States of America.

COMMITTEE ON SCIENTIFIC EVALUATION OF THE INTRODUCTION OF GENETICALLY MODIFIED MICROORGANISMS AND PLANTS INTO THE ENVIRONMENT

Steering Committee

Robert H. Burris (Chairman), University of Wisconsin, Madison
Fakhri A. Bazzaz, Harvard University, Cambridge, Massachusetts
Ralph W. F. Hardy, BioTechnica/Boyce Thompson, Ithaca, New York
Edward L. Korwek, Law Offices of Hogan & Hartson, Washington, D.C.
Richard E. Lenski, University of California, Irvine
Eugene W. Nester, University of Washington, Seattle
Stanley J. Peloquin, University of Wisconsin, Madison
Calvin O. Qualset, University of California, Davis
Ralph S. Wolfe, University of Illinois, Urbana

Subcommittee on Plants

Stanley J. Peloquin (Chairman), University of Wisconsin, Madison
Roger N. Beachy, Washington University, St. Louis, Missouri
Donald N. Duvick, Pioneer Hi-Bred International, Inc., Johnston, Iowa
Robert T. Fraley, Monsanto Company, St. Louis, Missouri
Ralph W. F. Hardy, BioTechnica/Boyce Thompson, Ithaca, New York
Richard N. Mack, Washington State University, Pullman
Ann M. Vidaver, University of Nebraska, Lincoln

SUBCOMMITTEE ON MICROORGANISMS

Richard E. Lenski (Chairman), University of California, Irvine
Peter J. Bottomley, Oregon State University, Corvallis
Ananda M. Chakrabarty, University of Illinois, Chicago
Rita R. Colwell, University of Maryland, College Park
Steve K. Farrand, University of Illinois, Urbana
Robert Haselkorn, University of Chicago
Roger D. Milkman, University of Iowa, Iowa City
Luis Sequeira, University of Wisconsin, Madison
James M. Tiedje, Michigan State University, East Lansing

Board On Biology Staff

Alvin G. Lazen, Study Director
Clifford Gabriel, Senior Staff Officer
Joseph L. Zelibor Jr., Staff Officer
Juliette Walker, Administrative Secretary
Kathy Marshall, Senior Secretary
Caitilin Gordon, Editor

BOARD ON BIOLOGY

COMMISSION ON LIFE SCIENCES

Preface

The potential benefits from the use of genetically modified microorganisms and plants are enormous—bacteria that biodegrade environmental pollutants, trees that grow more rapidly, food plants that flourish under saline or dry conditions, viruses that control insect pests, and a productive and economical agricultural enterprise whose plants use less fertilizer and resist pests. National, state, and local governments are considering safeguards to ensure that such benefits are maximized while possible hazards to the health and welfare of humans and damage to the environment are minimized. The necessity exists for timely field research of genetically modified microorganisms and plants in environments similar to those in which they eventually will be used. A flexible, well-reasoned, scientifically based oversight system must be applied so that tests of genetically modified organisms in the field can proceed when they are deemed safe.

Before deciding whether to allow a field test, we are first obliged to define what scientific information and issues must be considered, and then we must ask whether we know enough scientifically to be able to determine the relative safety or risk of the introduction. To obtain a reasoned consensus about these questions, the Biotechnology Science Coordinating Committee asked the National Academy of Sciences to prepare a report on the introduction of genetically modified plants and microorganisms into the environment up to the level of field testing.

A steering committee was formed by the National Research Council within the Board on Biology of the Commission on Life Sciences. The steering committee was responsible for the final report; two subcommittees also were organized, one concerned with plants and the other with microorganisms, each chaired by a member of the steering committee but with a membership that extended the expertise available to address the questions in a balanced fashion.

Over the past ten months, the committees have grappled with an extensive body of facts, concepts, and opinions and have attempted to formulate a concise, thoughtful, and objective summary that can contribute to resolving the issues. The committees felt their most important task was to reach a consensus about the science surrounding the issues of environmental introductions. Our committees had the luxury of doing this unbound by, but not oblivious to, the existing regulatory principles and approaches that have been applied in this area.

The report that follows starts in the middle in the sense that the committee has not attempted to write a primer on new technology, such as recombinant-DNA techniques, nor to provide a detailed background on the biological information that has led to our present level of knowledge. Many other sources provide such information. Rather, we have focused on the issues regarding planned testing of genetically modified plants and microorganisms in the laboratory and field.

As with all committee-written documents, not every member may agree with every statement. However, the report represents a consensus to which all members agreed.

The committee benefited from discussions with members of the Biotechnology Science Coordinating Committee and thanks them and its chairman, Dr. James Wyngaarden, for meeting with the committee and defining the charge.

Special thanks are due the chairmen of the subcommittees, Dr. Richard Lenski and Dr. Stanley Peloquin, who graciously accepted an enlarged share of the work load. All members of the committees deserve credit for hewing to the short deadlines. Timely completion of the report would have been impossible without their dedication and concern.

Reviewers of the report, though anonymous, deserve thanks for their important contribution; they have given close attention to the content, style, and scientific integrity of a report dealing with complex and sometimes contentious issues.

Staff members of the Board on Biology and the Commission on Life Sciences were invaluable in their assistance. Dr. John Burris provided wise counsel as the committee work proceeded. Dr. Alvin Lazen, study director, kept the committee and staff on the path in pursuit of a finished report. Dr. Clifford Gabriel and Dr. Joseph Zelibor, Jr., guided and coordinated the efforts of the subcommittees on plants and on microorganisms, respectively, and worked closely with the steering committee. Ms. Juliette Walker and Ms. Kathy Marshall skillfully and patiently arranged meetings and handled the administrative and clerical work. We thank Dr. Caitilin Gordon for her expert editorial assistance in the final stages of writing the report.

We sincerely hope that our report will help attain what we all seek—a safe and prudent use of a technology that holds tremendous promise for advancing the welfare of humanity.

Robert H. Burris
Chairman

Contents

1
Executive Summary

In late 1988, the Biotechnology Science Coordinating Committee (BSCC), representing the U.S. Department of Agriculture, Environmental Protection Agency, National Institutes of Health, National Science Foundation, and Food and Drug Administration, asked the National Academy of Sciences-National Research Council (NRC) to evaluate scientific information pertinent to making decisions about the introduction of genetically modified microorganisms and plants into the environment. The NRC was asked to use this analysis to identify criteria for defining risk categories and to recommend ways to assess the potential risks associated with introducing these modified organisms. A steering committee was formed under the Board on Biology of the NRC's Commission on Life Sciences to prepare a report responding to the BSCC request. The steering committee, with overall responsibility for preparing the report, was augmented by two subcommittees of experts, one for microorganisms and the other for plants.

The committee considered the foci of its work to be:

- plants and microorganisms, but not animals;
- introductions under field-test conditions typical of those currently being proposed, but not large-scale commercial applications and the scientific, economic, ethical, and societal issues associated with large-scale applications;

- environmental effects excluding human health effects;
- scientific issues primarily, not regulatory policy matters;
- field-test conditions only in the conterminous United States in recognition that domesticated and wild species are different in other countries and areas of the world;
- general procedures for determining categories of risk for introductions, not recommendations for specific cases.

The steering committee and subcommittees adopted the fundamental principle enunciated in the document "Introduction of Recombinant DNA-Engineered Organisms into the Environment: Key Issues" (NAS, 1987) that safety assessment of a recombinant DNA-modified organism "should be based on the nature of the organism and the environment into which it will be introduced, not on the method by which it was modified." The principle that evaluation should be of the product and not the process by which the product is obtained is reemphasized in Chapter 2 of this report. The discussion also points out that although genetic modification by molecular methods may be more powerful and capable of producing a wider range of phenotypes, "no conceptual distinction exists between genetic modification of plants and microorganisms by classical methods or by molecular methods that modify DNA and transfer genes."

The section of the report on plants (Chapters 3-6) discusses the relevant biological properties of genetically modified plants. It also describes past experience with genetic modification and introductions of plants modified by classical and by molecular genetic methods. The major environmental issue of potential weediness receives special attention in the report.

The section of the report on microorganisms (Chapters 7-11) discusses the properties of the genetic modification, phenotypic properties of the source organism and its genetically modified derivatives, and properties of the environment with respect to the organisms that may be released into it.

Investigators modifying microorganisms for environmental introduction should assess the influence of genetic alteration on the organism's phenotype and the mobility of the altered trait. It is highly unlikely that moving one or a few genes from a known pathogen to an unrelated nonpathogen will confer pathogenicity on the recipient. If the recipient is itself a pathogen, increased virulence for particular hosts may result. If modifications of this latter type are contemplated, special attention must be paid to them. In some cases persistence is not desirable and uncertainty exists about the microorganism's effects on the immediate environment. When assessing risk in these cases, the most relevant phenotypic properties relate to the persistence of the microorganism and its genetic modification. Evaluation of phenotypic properties raises questions about the fitness of the genetically modified microorganism, the potential for gene transfer from the introduced microorganism, the tolerance of the introduced microorganism to physicochemical stresses, its competitiveness, the range of substrates available to it and, if applicable, the pathogenicity, virulence, and host range of the introduced microorganism.

The report discusses the long history of utility and safety in the use of plants and microorganisms. Society has benefited greatly from the use of genetically modified microorganisms and plants, and field testing is essential if we are to increase our knowledge about the relative safety or risk of large-scale use of genetically modified organisms and to determine the potential utility of the modified organisms.

Other major scientific conclusions are as follows:

PLANTS

1. Plants modified by classical genetic methods are judged safe for field testing on the basis of experience with hundreds of millions of genotypes field-tested over decades. They are, in the terms used by the plant subcommittee, "manageable by accepted standards." The committee emphasizes that *the current means for making decisions about the introductions of classically bred plants are entirely appropriate and no additional oversight is needed or suggested in this report.*

2. Crops modified by molecular and cellular methods should pose risks no different from those modified by classical genetic methods for similar traits. As the molecular methods are more specific, users of these methods will be more certain about the traits they introduce into the plants. Traits that are unfamiliar in a specific plant will require careful evaluation in small-scale field tests where plants exhibiting undesirable phenotypes can be destroyed.

3. At this time, the potential for enhanced weediness is the major environmental risk perceived for introductions of genetically modified plants. The likelihood of enhanced weediness is low for genetically modified, highly domesticated crop plants, on the basis

of our knowledge of their morphology, reproductive systems, growth requirements, and unsuitability for self-perpetuation without human intervention.

4. Confinement is the primary condition for ensuring safety of field introductions of classically modified plants.

5. Depending on the crop species, proven confinement options include biological, chemical, physical, spatial, environmental, and temporal isolation, as well as size of field plot.

6. Plants grown within field confinement for experimental purposes rarely, if ever, escape to cause problems in the natural ecosystem.

7. Established confinement options are as applicable to field introductions of plants modified by molecular and cellular methods as to introductions of plants modified by classical genetic methods.

MICROORGANISMS

1. The precision of many of the molecular methods allows scientists to make genetic modifications in microbial strains that can be fully characterized, in some cases to the determination of specific alterations of bases in the DNA nucleotide sequence.

2. The molecular methods have great power because they enable scientists to isolate genes and to transfer them across biological barriers.

3. Although field experience provides considerable information about some microorganisms—for example, rhizobia, mycorrhizae, and many plant pathogens and biocontrol agents—in general, information regarding the ecology of microorganisms and experience with planned environmental introductions of genetically modified microorganisms is limited compared with that regarding plants. However, no adverse effects have developed from introductions of genetically modified microorganisms. Ecological uncertainties can be addressed scientifically with respect to genetic and phenotypic characterization of the microorganisms as well as by consideration of environmental attributes such as nutrient availability. Field tests of genetically modified organisms can go forward when sufficient information exists to permit evaluation of the relative safety of the test.

4. The likelihood of possible adverse effects can be minimized or eliminated by appropriate measures to confine the introduced microorganism to the target environment, for example, by introducing "suicide" genes, as they become practicable, into the organisms.

FRAMEWORK

The committee developed a framework for the evaluation of risk based on criteria that are summarized below and detailed in Chapters 6 and 11.

- Are we *familiar* with the properties of the organism and the environment into which it may be introduced?
- Can we *confine or control* the organism effectively?
- What are the probable *effects* on the environment should the introduced organism or a genetic trait persist longer than intended or spread to nontarget environments?

When the familiarity standard for a plant or microorganism has been satisfied such that reasonable assurance exists that the organism and the other conditions of an introduction are essentially similar to known introductions, and when these have proven to present negligible risk, the introduction is assumed to be suitable for field testing according to established practice.

The familiarity criterion is central to the suggested framework of evaluation. Its use permits decision-makers to draw on past experience with the introduction of plants and microorganisms into the environment, and it provides future flexibility. As field tests are performed, information will continue to accumulate about the organisms, their phenotypic expression, and their interactions with the environment. Eventually, as our knowledge increases, entire classes of introductions may become familiar enough to require minimal oversight.

Familiar does not necessarily mean safe. Rather, to be familiar with the elements of an introduction means to have enough information to be able to judge the introduction's safety *or* risk.

When knowledge of the type of modification, the species being modified, or the target environment is insufficient to meet the familiarity criteria, the proposed introduction must be evaluated with respect to the ability to confine or control the introduced organism and to the potential effects of a failure to confine or control it. The results of these latter evaluations will define the relative safety or risk of a proposed introduction.

The frameworks for microorganisms and plants differ in nomenclature and in emphasis on particular issues, mainly because of differences in life cycles, mechanisms of gene transfer, dispersal and containment or control procedures, persistence, and environmental factors. Fewer proposed field tests of microorganisms than plants may meet the familiarity criterion because the data base, from a history of planned introductions, is more limited at this time. Means to *confine* plants are well established and can be relatively simple, whereas means to *control* microorganisms appear to be more difficult. As a consequence, the subcommittee on microbiology suggests in its framework a close link between considerations of control and possible effects. The plant subcommittee's framework shows a distinct separation between considerations of confinement and of environmental effects.

We believe that our evaluation of the scientific issues and our proposed frameworks provide the responsible government agencies with the foundation for a flexible, scientifically based, decision-making process. Use of the frameworks for evaluation of field tests permits the classification of an introduced organism into a risk category.

2
Introduction

Recent advances in biology have proceeded at an astonishing rate, and biologists now have the means, by directly modifying genes, to alter living organisms more quickly and more precisely than has been done by nature and humans over millennia. There is general agreement that this ability can yield far-reaching improvements in our environment and in medical and agricultural practice. However, field testing of promising products of the new technology has been slowed by the absence of a full scientific consensus on the relative safety and risks of introducing modified organisms into the environment. Furthermore, the specific questions that are most important to consider in making decisions have not been agreed on. Hence, this NRC committee was formed to attempt to determine a reasoned consensus about what scientific questions must be asked and how such questions can aid in the development of a decision-making process based soundly on the facts of science.

The history of efforts to reach a common ground about the relative safety or hazard of genetic modification of organisms can be traced directly to the early 1970s, when advances in biological knowledge had given scientists the tools to recombine DNA in the laboratory into new sequences (see Appendix).

THE GENETIC MODIFICATION OF ORGANISMS: MERGING CLASSICAL AND MOLECULAR TECHNIQUES

This report describes the properties of plants, microorganisms, and the environment that must be evaluated when the introduction of a genetically modified organism into the environment is being planned. In this introductory section we explore the basic biological principles that underlie both classical and molecular means of altering the genetic makeup of organisms and explain how our interpretation of these principles leads to the conclusion that the products of classical and molecular methods are fundamentally similar. Both methods of modifying DNA produce an organism (product) that is genetically different from the starting organism regardless of the method (process) used. The molecular techniques are often more precise than classical techniques and can modify single nucleotides of bacterial genomes. Molecular modifications surpass classical techniques in their ability to introduce a great variety of traits from a wide range of donor organisms into the recipient organisms. As a corollary, the molecular techniques can generate a greater range of phenotypes than the classical methods. These principles as they apply to plants and microorganisms are discussed in greater detail in the sections of this report dedicated to the two kinds of organisms.

Plants and microorganisms contain nucleotides in combinations and arrangements that endow the organisms with genetic determinants for many traits. Other regions of DNA may control the expression of the traits. The DNA provides the raw material upon which genetic modifications depend. The evolution of new forms of crop plants and microorganisms results from selecting organisms with desirable traits from populations that possess heritable variation. When genetic variants are selected to produce the next generation, the population is changed with respect to the frequency of individuals having the selected characteristic. In the terms used in population genetics, selective breeding or propagation changes gene frequencies, and the population differs in some aspect from its predecessor even though the change may be small.

Modification of microorganisms and plants can be performed by either classical or molecular methods. No hard line exists between the two categories, especially with microorganisms. For this report, we generally include as classical those means of genetically modifying organisms that were used before recombinant DNA techniques were developed. One major distinction of classical methods is that they are relatively undirected modifications of the genome. Molecular methods provide more flexibility and control and thus are more specific in directing the modifications toward a planned end product.

Methodological and biological distinctions exist in culturing microorganisms and plants, but one feature of the new genetic technologies is that they permit us to manipulate plants at the cellular level. This technology provides new commonalities to plant and microbial breeding.

Classical methods are those in which the genetic recombinations occur essentially in a natural way; desirable offspring variants are then selected in the laboratory or the field. Examples include spontaneously mutating microorganisms and sexually cross-bred plants. The term classical also includes some methods called that only because they predate the introduction of modern gene-splicing techniques. The latter include such human-mediated techniques as exposure of organisms to chemical mutagens or physical agents such as x-rays and ultraviolet radiation. We also include as classical those mechanisms of DNA transfer that occur without chemical treatment of a cell's envelope, such as transformation, conjugation, and transduction in microorganisms.

Molecular methods of genetic modification include the newer methods for modifying DNA in which one nucleotide can be substituted for another at a predetermined site in a DNA molecule (site-directed mutagenesis). Molecular gene transfer methods are used for transfer of genetic material between donor and recipient cells that have diverged widely through evolution and probably do not exchange DNA without laboratory manipulation. However, it is important to recognize that certain gene transfers thought impossible in nature a few years ago because of the phylogenetic distance between donor and recipient have now been shown to occur in the laboratory, and they may occur in nature. For example, there is evidence that a gene or genes for erythromycin resistance was transferred between the gram-negative bacterium *Campylobacter* and unrelated gram-positive bacteria (Brisson-Noel et al., 1988). Recent laboratory experiments have accomplished gene transfer between *Escherichia coli* and streptomyces (Mazodier et al., 1989) or yeast (Heinemann and Sprague, 1989). Another example relates to the natural transfer of DNA from the bacterial species *Agrobacterium* to plant cells (Nester et al., 1984). Plasmid genes from this bacterium probably were transferred into a species of tobacco early in the evolution of the genus *Nicotiana*, and they became integrated into the plant chromosome. These genes, or their remnants, have been detected in a variety of different species of *Nicotiana*, which presumably evolved from the original infected plant (Furner et al., 1986).

PLANT MODIFICATIONS—CLASSICAL TECHNIQUES

Spontaneous and mutagen-induced variation in plants has produced a great variety of genetic traits that may be used in plant breeding. The crop plants of today had their origins in the fields of early farmers who selected plants with desirable traits and perpetuated plants to meet agricultural needs.

Controlled matings (hybridization) of plants through the sexual process is the cornerstone of classical plant breeding. Hybridization and selection of plants with new combinations of traits have been used to increase genetic diversity. By repeated hybridization and selection, new traits could be introduced into varieties already proven successful in agriculture.

Hybridization is often possible between species, usually within the same genus. However, many interspecific hybridizations require human-mediated intervention to facilitate the sexual process. For example, developing embryos are excised and cultured on nutrient media before being grown as plants in the field. The male or female fertility of such hybrids is often reduced so that they themselves must be hybridized with one of the parents or with a closely related species. Alternatively, fertility can be restored by doubling the chromosome number. With sexual hybridization, the resulting progeny contain full complements of genes from each parent. The challenge for plant breeders is to select for the genes which result in a plant's exhibiting the desired combination of traits. Because interspecific hybrids, and even many intraspecific hybrids, have a parent that may be poorly adapted to survive and grow in an agriculturally useful way, considerable effort is required to examine large numbers of plants to find the desired combinations of traits.

Two major limitations exist with classical plant breeding. The first is an extraordinarily large degree of variability from which a low frequency of desired plants must be identified. Second, the gene pool—the source of genes accessible to the breeder—is limited to those species which can be sexually hybridized.

PLANT MODIFICATIONS—MOLECULAR TECHNIQUES

In principle, any gene can now be introduced into any plant by one of several possible molecular modification techniques. At present, the most frequently used agent for DNA transfer is the common soil bacterium *Agrobacterium* (Nester et al., 1984). This organism evolved a mechanism for transferring part of its plasmid into plant cells, where it is integrated randomly into the chromosome (Peerbolte et al., 1986). The introduced DNA is inserted within this plasmid DNA as a "hitchhiker." Once integrated into the plant's chromosome, the DNA is transmitted from parent to offspring and follows the pattern of Mendelian inheritance. Virtually all dicotyledonous plants are amenable to transformation by *Agrobacterium*, but most monocotyledonous plants appear to be resistant.

A technique frequently used to transform monocotyledonous plants, such as maize and rice, is electroporation; this technique requires removal of the plant cell walls before the DNA is added. These naked cells, or protoplasts, often do not synthesize new cell walls readily. Thus, regeneration of whole, fertile plants from protoplasts has limited use for molecular gene transfer, especially in cereal grasses. More recently, DNA-coated gold or tungsten particles have been "shot" into plant cells, and stable, genetically transformed plants have been regenerated from the cells or organized tissue (Klein et al., 1987). This technique may be suitable for introducing DNA into plant chloroplasts (Boynton et al., 1988) and mitochondria (Johnston et al., 1988), as well as into the nucleus. Current research is directed toward introducing DNA into specific plant tissues that have the greatest probability of regenerating genetically modified plants.

COMPARISON OF CLASSICAL AND MOLECULAR TECHNIQUES IN PLANTS

The major difference between classical and molecular techniques is the greater diversity of genes that can be introduced by molecular techniques and the greater precision of these introductions. From a single gene to more than 50 genes can be introduced with the *Agrobacterium* system, although the site in the plant chromosome at which the foreign DNA has been integrated appears to be random. The donor DNA can be derived from the same or different plant species, or even from microorganisms or animal cells. For example, the DNA from fireflies (Ow et al., 1986) and bacteria (Koncz et al., 1987) that codes for luminescence has been inserted into plants. Thus, no species barrier exists, because the chemical nature of DNA is universal in its structure, irrespective of the organism of its origin.

After being integrated, the gene, to be useful, must be expressed in the host plant. Genes have regions at one end of their nucleotide chain that control when and under what conditions the gene will be expressed. These regions determine specific conditions for gene expression, for example, in the light, in specific tissues, or at certain stages of development (Goldberg et al., 1989). On the basis of this knowledge and recombinant DNA technology, one can attach the desired region of a gene to a bacterial gene and introduce the combination into a plant cell, where it will be expressed in a specific tissue. Particular conditions, such as wounding, may be needed for expression of the added gene or genes, and knowledge of these conditions can be used to precisely control expression (Ryan, 1988).

GENOME MODIFICATION OF MICROORGANISMS—CLASSICAL TECHNIQUES

The classical methods of genome modification in microorganisms fall into two classes, selection of spontaneous and induced mutations and the exchange of DNA between (usually) closely related organisms. Spontaneous mutations result in a variety of heritable changes in the DNA, including the substitution of one nucleotide for another, the deletion or addition of one or more nucleotides, and other types of DNA rearrangements. Many spontaneous mutants appear to result from the movement of transposable elements to new locations in the cell's DNA. Transposable elements, first discovered in maize, also occur in other plants (McClintock, 1950), bacteria, and animals.

Another mechanism for generating variability in microorganisms is through the introduction of new genetic information from either chromosomal or plasmid DNA. DNA from a donor organism's chromosome is integrated into the recipient genome. Plasmids, being selfreplicating, do not have to integrate their DNA into the genome of the recipient. Consequently, plasmid DNA can be transferred to more widely divergent organisms than DNA from the chromosome of a donor organism. Plasmid movement can be monitored because the DNA often provides the genetic code for readily distinguishable traits, such as antibiotic resistance.

In bacteria, gene transfer can occur by three different classical means: DNA-mediated transformation, in which the DNA is transferred as "naked" DNA; transduction, in which the DNA is enclosed in a virus coat and the virus mediates the transfer; and conjugation, in which the DNA is transferred during cell-to-cell contact between donor and recipient cells. Presumably, all these mechanisms operate in nature (Freifelder, 1987).

GENOME MODIFICATION OF MICROORGANISMS—MOLECULAR TECHNIQUES

The range of techniques to mutate bacteria has expanded and become sophisticated in recent years. It now is routine practice to mutate specific genes (insertion mutagenesis) (Ruvken and Ausubel, 1981) as well as to alter specific nucleotides within a gene (site-directed mutagenesis) (Kunkel, 1985). These techniques are possible not only for microbial genes, but, in principle, for genes from any organism.

The range of microorganisms among which DNA can be transferred has also been expanded through the use of new technologies. Thus, it is now possible to transform cells by physically altering their cell envelopes so that they become permeable to most DNA molecules. One such technique is electroporation, in which recipient cells and the genetic material to be transferred are subjected to an electric current (Fromm et al., 1987). The successful use of these techniques for genome modification requires that the entering DNA be able to replicate inside its new host. In principle, the techniques for performing these manipulations are straightforward. With such techniques, plasmids have been constructed that can replicate in both the bacterium *E. coli* and the yeast *Saccharomyces cerevisiae* (Freifelder, 1987).

COMPARISON OF CLASSICAL AND MOLECULAR TECHNIQUES IN MICROORGANISMS

Recent molecular technological advances in mutagenesis and gene-transfer methods have opened new possibilities for expanding the range of microorganisms into which DNA from unrelated organisms can be introduced. The genus barrier and, indeed, the kingdom barrier are no longer complete obstacles.

Recombinant DNA methodology makes it possible to introduce pieces of DNA, consisting of either single or multiple genes, that can be defined in function and even in nucleotide sequence. With classical techniques of gene transfer, a variable number of genes can be transferred, the number depending on the mechanism of transfer; but predicting the precise number or the traits that have been transferred is difficult, and we cannot always predict the phenotypic expression that will result. With organisms modified by molecular methods; we are in a better, if not perfect, position to predict the phenotypic expression.

With classical methods of mutagenesis, chemical mutagens such as alkylating agents modify DNA in essentially random ways; it is not possible to direct a mutation to specific genes, much less to specific sites within a gene. Indeed, one common alkylating agent alters a number of different genes simultaneously. These mutations can go unnoticed unless they produce phenotypic changes that make them detectable in their environments. Many mutations go undetected until the organisms are grown under conditions that support expression of the mutation.

SUMMARY

We have reviewed briefly the various means by which plants and microorganisms can be genetically modified by methods termed "classical" or "molecular." Genetic variability in microorganisms and plants is enhanced by classical modifications such as spontaneous or mutagen-induced variation, by hybridization, and by gene transfer. These methods are relatively imprecise and undirected and less powerful than molecular techniques for modifying genes. However, no conceptual distinction exists between genetic modification of plants and microorganisms by classical methods or by molecular techniques that modify DNA and transfer genes.

Figure 2-1 graphically depicts this view. The difference in the modes of genetic modification are not deemed critical, and both methods are included in one box. This figure also illustrates that no distinction exists between so-called classical and molecular breeding methods at the steps of evaluation in laboratory, field, or large-scale environmental introduction.

This understanding of the biological principles has the following implications for the report:

1. The deliberations of the committees were guided by the conclusion (NAS, 1987) that the *product* of genetic modification and selection should be the primary focus for making decisions about the environmental introduction of a plant or microorganism and not the *process* by which the products were obtained.
2. Information about the process used to produce a genetically modified organism is important in understanding the characteristics

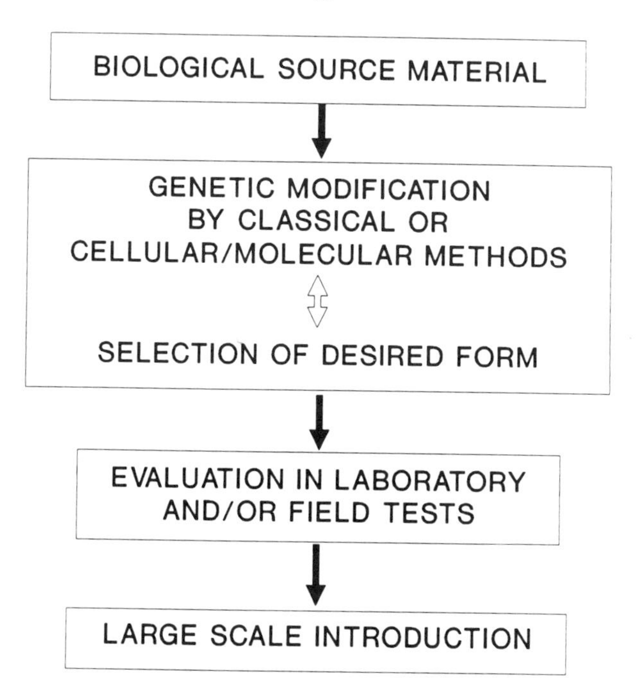

FIGURE 2.1 Genetic modification of an organism and its introduction into the environment.

of the product. However, the nature of the process is not a useful criterion for determining whether the product requires less or more oversight.

3. The same physical and biological laws govern the response of organisms modified by modern molecular and cellular methods and those produced by classical methods. Scientists have vast experience with the products of classical modification, and the knowledge gained thereby is directly applicable to understanding, evaluation, and decision-making about the relative safety or risk of field tests on products of molecular modification techniques.

3
Past Experience with Genetic Modification of Plants and Their Introduction into the Environment

For thousands of years, plants have been improved by genetic modification. Ancient agriculturists selected plants with desirable traits from landraces of domesticated relatives of wild species. Landrace populations consist of mixtures of genetically different plants, all of which are reasonably adapted to the region in which they evolved but differ in many characteristics including reaction to disease and insect pests. With the rediscovery in 1900 of Mendel's concepts of inheritance, the scientific application of genetic principles to crop improvement began. Each scientific advance has increased our ability to alter the genetic makeup of plants predictably, and several techniques are often used together to improve plants. For example, an existing plant chosen for genetic modification by recombinant DNA techniques might have been modified by many generations of classical breeding and selection; the recombinant plant derived from the original could then be reintroduced into a classical breeding program from which its descendants would be released for commercial use. Each technique for genetic modification constitutes only one component in the entire crop-improvement process. Figure 3-1 indicates the sequence of scientific advances that has given us our present ability to modify plant genomes in ways and at a pace heretofore impossible. The basic goal of improving crops and other plants, which is still being pursued actively, includes improvement of agronomic traits, crop-end-use quality, and pest resistance.

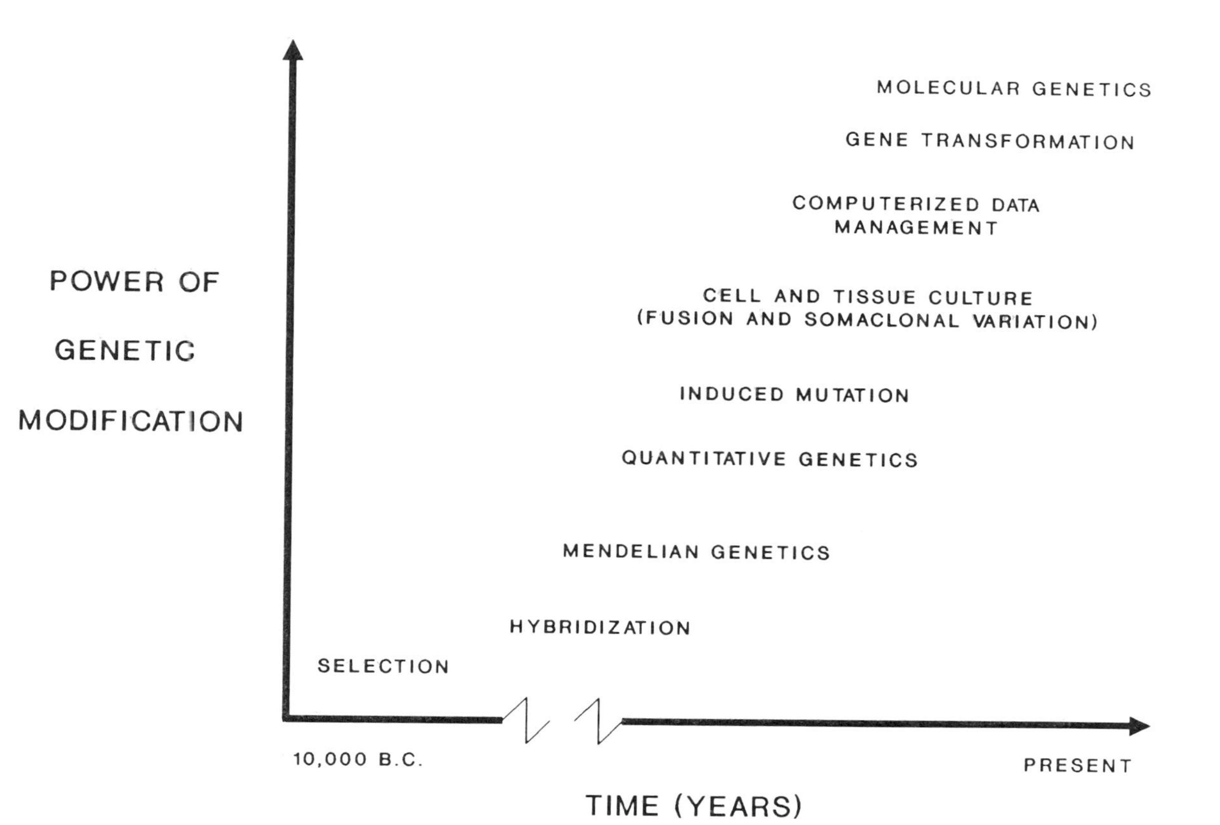

FIGURE 3.1 Increase in power of genetic modification over time.

In this chapter we place the remaining chapters in perspective. This chapter includes discussions of the classification of various techniques for genetic modification, the results of genetic modifications, case studies of field introductions of crop plants, and our experience with confinement methods.

TYPES OF GENETIC MODIFICATION IN PLANTS

The many techniques available to modify plants genetically can be divided into three main categories: classical, cellular, and molecular. Each of these results in genetic variation, but each provides a different avenue for producing a plant with desirable traits.

Classical Techniques of Genetic Modification of Plants

Hybridization. Most genetic modification techniques are used by plant breeders whose purpose is to apply the techniques to improve plants with commercial value. Historically, breeders have been limited by the natural or induced sexual compatibility of plants to be hybridized in their crop-improvement programs. However, new techniques, such as molecular techniques for genetic modification, are used in crop- and other plant-improvement programs to bypass the sexual hybridization step. These newer techniques complement those of classical plant hybridization.

Undirected Mutagenesis. Mutations can be induced in the DNA of plant cells by such techniques as the use of DNA-altering chemicals or ionizing radiation (x-rays). Intact plants or plant cells are treated with the mutagenic agent and then selected for desirable traits. This process is random, and it can induce undesirable as well as desirable changes. Mutagenesis has been used effectively to generate agriculturally important traits (Konzak et al., 1984). Although the range of useful variations has been narrow, more than 150 plant varieties bearing traits induced by mutagenesis have been released.

Anther and Ovule Culture. In plant breeding and in other plant research, it is sometimes desirable to have plants with half the original number of chromosomes. If a plant is diploid (2x), haploid (1x) gametes or cells found in the anthers and ovules can be cultured to produce haploid plants. These genetically modified plants can then be used in breeding or in basic research. Anther and ovule culture used for obtaining haploids is followed by chromosome doubling to give homozygous diploid plants for use as cultivars or as parents of hybrids (Chase, 1969).

Embryo Rescue. Embryo rescue or culture is a procedure whereby a sexual cross yielding a viable embryo but abnormal endosperm is "rescued" by culturing the embryo from the nonviable seed to produce a mature plant. This cultured plant can be used in further breeding; for example, the procedure has been used as an integral part of producing barley varieties (Choo et al., 1985).

Cellular Techniques of Genetic Modification of Plants

Somaclonal Variation. Somaclonal variation occurs in plants regenerated from cell in tissue culture, presumably as a result of stress imposed on the plant cells. The genetic changes underlying somaclonal variation include whole chromosome changes, small and large deletions and chromosome rearrangements, single base changes, and insertion mutations resulting from the activation of cryptic transposable elements (Orton, 1983; Vasil, 1986).

Cell Fusion. As in sexual hybridization in breeding, cell-fusion techniques recombine plant genomes. Cell fusion is especially useful with plants not fully sexually compatible. The cells are dissociated from tissues, walls are stripped from the cells, the membranes of the resulting protoplast are modified to facilitate fusion, and after fusion the protoplasts are cultured and regenerated into intact plants. This technique can produce novel combinations of nuclei, mitochondria, and chloroplasts (Ehlenfeldt and Helgeson, 1987).

Molecular Techniques of Genetic Modification of Plants

Molecular techniques offer several advantages and complement existing breeding efforts by increasing the diversity of genes and germ plasm available for incorporation into crops and by shortening the time period for commercial release. The many molecular techniques for genetic modification of plants can be divided into two main types: vectored and nonvectored. These techniques are discussed in detail in Chapter 5.

Vectored Modifications. Vectored modifications rely on the use of biologically active agents, such as plasmids and viruses, that facilitate the entry of the foreign gene into the plant cell.

Nonvectored Modifications. Nonvectored modifications rely on the foreign genes being physically inserted into the plant cell by such methods as electroporation, microinjection, or particle guns.

THE RESULTS OF GENETIC MODIFICATION

Plant breeding has sought to make two major kinds of modifications in recipient organisms: those to increase yield and those to increase reliability of performance.

Increased Yield and Increased Reliability of Performance

Maize breeders have looked for varieties or hybrids that produce larger amounts of grain per unit of land area, potato breeders for increased tuber yields, and cotton breeders for increased yields of lint (fiber). In addition to breeding for greater yield one may breed for a product with more desirable qualities. Breeders of bread wheats, for example, must combine selection for maximum yield with selection for an optimal balance of the endosperm proteins required for good bread-making. Cotton breeders must select for maximum yield of fiber that also has desirable spinning characteristics.

The second obligation of plant breeders has been to select for reliability of performance. Components of reliability include resistance to diseases and pests as well as with the physical environment. Varieties that produce bumper yields in favorable growing seasons but fail to produce a crop in unfavorable seasons cannot be accepted by subsistence farmers. Their livelihood each year depends on the crops produced in the previous year. Commercial farmers in today's industrial nations have a less stringent requirement for reliability because storage facilities, crop insurance, and government subsidies reduce some of the problems caused by seasonal inconsistencies in production. But in the long run, commercial farmers need reliability of performance as well. Thus, plant breeders select for reliable varieties able to produce high yields of good quality.

Changes in Plant Architecture. Plant breeders, in modifying plant varieties, have selected them for their ability to produce changed and often highly unbalanced proportions of seeds, tubers, leaves, or whichever specific plant part is of economic or aesthetic interest. Genetically modifying an organism to increase the proportion of a specific plant part nearly always reduces the ability of the organism to maintain itself in the wild. Maize is one of the best known examples of a highly productive cultivated plant that cannot reproduce itself without human assistance. Its large, naked seeds bound together in a large ear and having no dispersal mechanism are notoriously ill-adapted for survival in the wild.

Changes in Pest and Disease Resistance. Plant varieties have been continually selected for improved resistance or tolerance to external factors that inhibit their inherent productivity. They have been selected for resistance to insect pests, to disease organisms, and, in recent years, even to specific herbicides. If such improved cultivars were also able to persist in the wild, they presumably would be better adapted (at least in the short term) to persist in the presence of disease, insects, and herbicides.

Improved Tolerance to Environmental Stresses. Cultivated plant varieties have also been selected through the years for better tolerance of environmental constraints to growth. Improvements are made in, for example, heat and drought tolerance, ability to withstand high moisture, tolerance of cold, ability to withstand excessive salts or high aluminum content in soils, ability to withstand iron deficiency induced by excessive alkalinity, and ability to prevail in competition with weeds through quick germination and extremely rapid growth in the seedling stage. If such improved cultivars persisted in the wild, they presumably would be better adapted to survive in the presence of a number of environmental constraints to growth. Breeders have a long history of incorporating these types of traits into crops without any evidence of enhanced weediness.

MODIFICATIONS AND THEIR EFFECTS ON PERSISTENCE

Although domesticated plants in general cannot survive and reproduce unless aided by humans, different degrees of survivability are found among different crops and at various levels of domestication within a crop. Further, genes from domesticated plants can potentially be transferred in pollen from these plants to their wild relatives. Thus, whether a cultivated crop is closely related to indigenous wild relatives is a factor that can affect survival of at least some of the genes or gene linkage blocks of domesticated plants.

Degree of Domestication

Maize has been cited as a cultigen so highly domesticated that it cannot survive and spread on its own. At the other extreme is a crop like *Cuphea*, just now being domesticated for use as an oilseed crop. Breeders have not been able to alter *Cuphea*'s self-sowing nature—the seeds drop from the plant at maturity, as in the wild species (Knapp, 1988). Thus, cultivated *Cuphea* could easily revert to the self-perpetuating nature of the wild species if other plant traits have not been altered by domestication to hinder survivability.

Most of the widely grown grain crops and the horticultural and vegetable crops are at the maize end of the reproductive spectrum; they cannot survive in the wild. Many of the forage and pasture crops—alfalfa, cool-season and warm-season grasses—cluster nearer the other end; they can persist with some degree of success or even total success. Each crop needs to be considered on its own capabilities for persistence and self-reproduction. Both the level of domestication and the reproductive phenotype of the plant must be considered. Thus, a highly selected hay or pasture crop, well-suited for farming needs as a forage plant, may be virtually unselected for any change in its seed dispersal mechanisms or in the ability of its seed to survive and give viable seedlings in the wild. Most alfalfa varieties, for example, still have a strong tendency to produce seed in dehiscent (self-sowing) pods, and seed dormancy may allow it to lie in the ground for years before germinating. Selection in alfalfa has been primarily for disease resistance and altered plant habit—for changing the phenotype of stem and leaf—not for altered reproductive structures.

Plant Habit

Plant architecture has a great effect on persistence and reproduction. The bush nature of the common garden bean greatly limits its adaptability; the wild bean in Mexico is a climbing vine, well-suited to survival by climbing up to sun and air on stems of sturdy tall grasses such as teosinte. In contrast, selections of Indian grass (*Sorghastrum nutans*), a highly vigorous and desirable United States warm-season pasture grass, are unchanged in plant phenotype from their wild prairie progenitor. These cultigens might be more competitive than their unselected progenitors if they were introduced back into native prairie ecosystems since they have been selected primarily for vegetative vigor.

Grain, vegetable, and fruit crops are generally selected for highly modified plant habit or fruit type that would not be favorable to persistence in the wild state; forage and pasture crops tend to differ less from wild relatives, but even they may have a more upright plant habit and faster growth rate. Such changes might place them at competitive disadvantage over time in the struggle for survival in the wild.

Adaptability, Range of Habitats

Survivability in the wild can be a broad-ranging but ill-defined term. The wild environment can refer (1) to pristine natural stands of vegetation essentially unaltered by humans or (2) to untended vegetation that is nevertheless altered by human activity because of such practices as lumbering, slash-and-burn agriculture, pasturing, or incidental traffic. Or the term can refer simply (3) to survival of "wild" plants—weeds—in cultivated fields. In general, domesticated plants have closest affinities to wild plants adapted to growth in periodically disturbed habitats. One theory contends that most domesticated plants were selected from the class of plants we now call weeds—plants well adapted to be pioneers, that is, rapid invaders of patches of ground laid bare by natural phenomena such as wind, fire, or flood (Anderson 1952). Humans with hoes, spades, and fire reproduced nature's open spaces in order to aid or ensure the growth of certain desired species already adapted to such conditions. Other unwanted pioneer species were thereby encouraged unintentionally, and came to be known as weeds.

Domesticated plants and their weeds have thus evolved together, and distinctions between them are sometimes minor. For example, grassy annual sorghums, grown as pasture crops or for cutting as green forage, have often retained their wild ancestors' traits of bearing self-sowing, long-lived seeds with varying periods of dormancy. Thus, they are adapted to selection for survival and reproduction as weeds in row-crops such as maize, where they can grow to maturity. Such revertant forage sorghums [known to farmers as shatter-cane, (Chapter 4)] have a further preadaptation to the modern chemical age. They have the same general pattern of herbicide resistance as maize (a fairly close relative taxonomically) and so are not controlled by most corn-field herbicides. Shatter-cane, in areas like Nebraska where a typical rotation is maize to sorghum, has become a weed;

it is controlled through the use of herbicides, cultivation, and crop rotation (Nilson et al., 1988).

Thus, range of adaptation to soil, water, climate, and chemicals is important in determining possible persistence of a cultigen.

CASE STUDIES OF INTRODUCED CROPS

When exotic plant species (wild or domesticated) are introduced into a new geographic location, their adaptability is uncertain. The vast majority of introduced species fail to establish populations that result in significant environmental harm (Simberloff, 1985). Most crop introductions (domesticated exotic species, such as soybean) have provided a large societal benefit and have caused either no or only very localized problems. A few plant introductions (usually exotic species, such as kudzu) have established themselves as weeds.

The vast majority of the crop plants grown in the United States have foreign origins. Only a small number of crops including sunflower, cranberry, Jerusalem artichoke, blueberry, and strawberry originated here. The bulk of the agricultural production in the United States has depended on the introduction of exotic species such as wheat, soybeans, peaches, cherries, apples, tomatoes, potatoes, and peas. This can be an inconvenience for breeders, because the useful gene pool found in wild relatives may be less readily accessible. This also can be an advantage, as genes introduced into these crop plants are not likely to spread to wild weedy populations because the growing area does not harbor native cross-hybridizing species. Instances in which introduced crops have escaped cultivation and have become localized weed problems are rare (see Chapter 4).

Soybean

The genus *Glycine* can be divided into two subgenera, which appear to have different geographic origins. The subgenus *Glycine* is distributed predominantly in Australia, and the subgenus *Soja* primarily in China and adjacent areas. The cultigen (cultivated soybean), *Glycine max* (L.) Merr., is in the subgenus *Soja* and originated genetically in China. The gene pool for the cultigen is limited to its relatives in the subgenus *Soja*, as only limited success has been achieved in hybridizing the cultigen with species in the subgenus *Glycine* (Hymowitz and Newell, 1981).

Between 1765 and 1898, the soybean was introduced into the

United States on many occasions and was grown both in small plantings and commercially for hay and as a forage crop. In 1898, only about eight cultivars were grown in the United States. However, a 1928 collecting trip to Japan, Korea, and northeast China brought back 4,451 new accessions to the United States (Hymowitz, 1984). Evaluated in field plantings throughout the country, these acquisitions contained a high degree of genetic variability that would be useful to breeders; for example, the genes carried resistance to many damaging diseases, such as brown spot, purple seed stain, *Phytophthora* root rot, soybean mosaic, and root-knot nematode (Hymowitz, 1984).

The soybean has been genetically modified with *Agrobacterium*-based transformation techniques (Hinchee et al., 1988) and with particle-gun technology (McCabe et al., 1988). These methods stably integrated the DNA in the soybean chromosomes. These methods have produced herbicide-tolerant soybeans, and field tests are being planted in the United States in 1989.

Extensive breeding programs have allowed the United States to become a world leader, producing 56 percent of the world's soybeans in 1985 (Hymowitz, 1987). Soybeans are grown on about 65 million acres of farm land annually in this country (USDA, 1986) and are a vital part of the nation's farm economy.

Canola

Canola is the general term for rapeseed in the genus *Brassica* developed by Canadian plant breeders in the 1950s to 1980s (Downey and Rakow, 1987). Historically rapeseed oil has been used as a lubricant and as an edible oil. The need for marine lubricating oils during the Second World War motivated Canadian farmers to initiate commercial growing of rapeseed, but the need disappeared after the war and production declined. Experiments in the 1940s and 1950s demonstrated that erucic acid, one of the major fatty acids in rapeseed oil, is metabolized poorly by mammals. In addition, erucic acid, when fed to test animals in sufficient quantities, was shown to induce heart lesions. Another drawback was that the meal recovered after oil extraction was limited as feed for nonruminant animals because of its high level of glucosinolates, compounds that release goiterogenic agents after enzymatic hydrolysis.

By classical plant-breeding methods, Canadian scientists selected variants and produced varieties with low concentrations of erucic acid in rapeseed oil (called LEAR oil), and they were released for commercial production in the late 1960s. A Polish cultivar of *Brassica napus* was identified with low glucosinolates, and this characteristic was rapidly introduced into LEAR. "Double-low" rapeseed varieties (low in erucic acid and glucosinolates) were released in 1974 in Canada and are now being introduced into Europe. The acreage of rapeseed in Canada increased sharply with each of the above developments.

Rapeseed, including canola, is sensitive to herbicides, making weed control difficult. In addition, atrazine soil residues make it difficult to grow rapeseed in fields treated with atrazine. In the late 1970s and early 1980s, plant breeders incorporated atrazine resistance from certain native *Brassica* weedy species into canola. A 20 percent reduction in yield is associated with herbicide resistance; however, more recent atrazine-resistant canolas show less yield penalty. Using molecular techniques, scientists have now produced a glyphosate-tolerant canola that has been field-tested in Canada (R. K. Downey, Agriculture Canada, personal communication, 1989).

The double-low *Brassica napus* and *B. campestris* varieties were the first rapeseed to meet specific quality requirements of low erucic acid and low glucosinolates. Rapeseed oil must contain less than 2 percent erucic acid, and the solid component of the seed must contain less than 30 micromoles of glucosinolate per gram to be classified as canola. Canola is now being adopted as a crop internationally. Canola oil was designated GRAS (generally regarded as safe) in the United States—as LEAR oil in 1985 and as canola oil in 1988. Canola oil has become the major edible oil in Canada, and its use worldwide is growing. Oilseed rape can be transformed by *Agrobacterium* vectors (Fry et al., 1987) and may represent one of the first crops in which herbicide- and disease-resistant plants produced by molecular modification are commercialized.

Potato

The early stages of domestication of the potato occurred about 8,000 years ago in the altiplano region of the border between Peru and Bolivia. It first appeared in Europe during the latter sixteenth century (about 1570 in Spain and 1590 in England). Potatoes were introduced into Germany, Poland, and Russia by the end of the seventeenth century and were of great commercial importance by the second half of the eighteenth century. They were brought from England and Ireland to North America between 1620 and 1680 (Hawkes, 1982).

The potato is unexcelled among cultivated plants in the abundance of related germplasm and the ease of incorporating this germplasm into cultivated forms. About 180 tuber-bearing wild species and several primitive cultivated species are known. They are distributed from the southern United States to southern Chile, with the largest number of species in the Andean regions of Peru and Bolivia. Potatoes occur from sea level to an elevation of more than 4,000 meters and in nearly every type of ecological location. They represent a polyploid series from diploids to hexaploids (Hawkes, 1982). Most important, the primitive cultivated and wild species are indispensable sources of resistance to diseases, pests, frost, and drought as well as sources of valuable processing characteristics. They also represent significant genetic diversity for breeding for heterotic (highly heterozygous) genotypes.

Resistance to viruses, bacteria, fungi, nematodes, and insects has been identified in primitive cultivated and wild species. Resistance has been successfully incorporated into useful cultivars by hybridization and selection. The extensive efforts to breed for disease and pest resistance, particularly in Europe, have led to the incorporation of germplasm from several species into many cultivars. The majority of cultivars in Europe and North America contain germplasm of from one to six species. Genes of *Solanum demissum* (a hexaploid species from Mexico with blight resistance) are incorporated into more than 50 percent of all cultivars. The genetic diversity provided by *S. demissum* benefits yield (Ross, 1986).

It has been possible to hybridize almost all wild species to the common cultivated potato either directly or indirectly by use of multiple crosses. Through several backcrosses of hybrids to existing cultivars, new, acceptable cultivars were obtained that contain the desired germplasm from the wild species. No undesirable "wild" trait has been observed that has not disappeared during this procedure.

From the time of early domestication of the potato to the present, thousands of cultivars have been bred and released, and several hundred of these have been grown on large acreages. Other plant species and the environment have apparently suffered adverse effects. One cultivar, found to have unsafe levels of particular alkaloids in the tubers, was withdrawn from the market. Advanced selections are now required for alkaloids to be tested before they are released (as required by GRAS) to ensure their acceptability in this regard. The variety with high alkaloids is the parent of several other important varieties, all of which have low levels of alkaloids themselves.

The potato is a favorable organism for cellular and molecular manipulations for two reasons: (1) Plants can be regenerated from protoplasts, leaf cell clusters, calli, and organized tissues such as stem apical meristems, and (2) *Agrobacterium* Ti-plasmids can be used for transformation (Fraley et al., 1986; Ooms et al., 1987). The direct transfer of genes for resistance into highly developed cultivars with gene-transfer methods would be significantly more effective than if done by classical breeding.

Potato plants regenerated from protoplasts or other unorganized groups of cells display an outburst of phenotypic variation. Some of this somaclonal variation is due to chromosomal changes, but the basis of other variation is not known. However, the somaclonal variants resemble the variants found in progeny from sexual crosses. Somaclones with an improved specific trait have been identified, although their overall performance has not been superior to the parental clone (Ross, 1986).

Somatic hybrids have been generated from both intraspecific and interspecific cell fusions. Many fusion hybrids between 24-chromosome *Solanum tuberosum* clones and the sexually incompatible, wild non-tuber-bearing species *Solanum brevidens* have been produced. These hybrids are of particular interest, since some are resistant to potato leaf roll virus (Austin et al., 1985; Gibson et al., 1988). Although chromosome number varied among the hybrids, several had the expected 48 chromosomes. Further, these hybrids can be hybridized to cultivars to obtain progeny for further selection and evaluation. A wide range of phenotypic variation among the somatic fusion hybrids resembled the somaclonal variation found in plants regenerated from protoplasts. Through special crosses, germplasm of *S. brevidens* can be incorporated into *S. tuberosum* by sexual crosses. The products of cell fusion are phenotypically similar to those of these sexual crosses.

Maize (Corn)

Introduction of new maize varieties into new environments probably has occurred since maize was first domesticated in Mexico, several thousand years ago. Maize entered North America several hundred years ago, constantly selected by Native Americans to allow its adaptation to northern climes and varying disease and weather problems. Only inferences, based on the archaeological record, suggest actual events in early times. However, events of the past 50 years are reasonably well detailed and documented.

Since the 1930s, maize breeders have relied on sexual crossing of elite, highly developed breeding lines followed by genetic recombination during several generations of self-pollination to develop new inbred lines that are suitable parents of commercial maize hybrids. The next step, yield testing for a 3- to 5-year period in both small plots and on those as large as farms, is crucial to developing seed products and to identifying new commercial hybrids with stable performance across a number of growing environments.

Gene flow from commercial maize varieties to the closely related teosintes in Mexico has been studied (Smith et al., 1981). Annual teosintes (closely related to maize, and also considered interfertile with it) exist in Mexico as weeds in corn fields and as completely wild species. For thousands of years, farmers in Mexico have been selecting specific new varieties of maize and reproducing them under conditions that allow the maize pollen to fall freely on stigmas of teosinte plants growing in the maize fields or nearby. Thus, there has been ample opportunity for the farmers' "deliberate release" to spread maize genes into the teosinte populations. Maize is notorious for being unable to persist in the wild because its seeds are unprotected and are tightly bound together in large ears, thus preventing their dispersal. Contamination of teosintes with maize genes for these traits would decrease the ability of the teosintes to persist in the wild. Nevertheless, various types of teosinte have maintained their distinctive phenotypes and their ability to reproduce and persist in the wild (Doebley, 1984). Biologists believe that there is limited gene flow from maize to the teosintes (and from teosintes to maize), but such gene flow does not seem to be detrimental to the teosintes nor to change their basic nature as distinctive wild races and species.

For decades, corn breeders have been modifying the corn genome by conventional breeding methods. Two situations are discussed here to exemplify the type of problems that have developed and how they have been readily managed by plant breeders.

The first example is breeding for resistance to northern corn leaf blight fungus (*Helminthosporium turcica*). A major gene for resistance to northern corn leaf blight, called *Ht1*, was introduced from two sources into U.S. corn-belt breeding populations about 25 years ago. It was bred into important inbred lines and widely used in hybrids. For many years the gene provided useful degrees of tolerance to northern corn leaf blight. Recent years have seen the appearance of a new biotype of the disease organism that flourishes in maize plants containing the *Ht1* gene. Thus the protection afforded by *Ht1* against the disease was greatly reduced. Because U.S. maize breeders had routinely and continually bred with non-*Ht1* sources of resistance to northern corn leaf blight, new hybrids were available immediately to substitute for those that suffered from the new race of northern corn leaf blight (D. N. Duvick, Pioneer Hi-Bred International, Inc., personal communication, 1989).

The second example is that of the southern corn leaf blight epidemic. In 1970, approximately 15 percent of the U.S. corn crop was destroyed by the fungal plant pathogen *Helminthosporium maydis*, which causes southern corn leaf blight (Zadoks and Schein, 1979). This represented a loss of 20 million metric tons of corn, worth about one billion dollars. Southern corn leaf blight was not a new corn disease, but, rather, one that had been controlled successfully with a variety of resistance genes. What then could account for the problem in 1970? Two key factors were involved: the natural development of a new race of the pathogen, race T, and the extensive use of hybrid lines with Texas cytoplasmic male sterility, T_{cms}.

The first factor to consider is the development of *Helminthosporium maydis* race T. Plant pathogens are continually evolving in response to selective pressures from changes in their environment, such as the introduction of new types of host plant resistance genes. This usually yields a number of different races that may be isolated geographically or biologically on more suitable alternative host plants. This was the situation for the southern corn leaf blight fungus. After examining collections of *H. maydis*, it was determined that race T was present in many parts of the world some 7 to 15 years before the 1970 epidemic. However, the fungus existed mainly on gramineous hosts and not on corn because commonly planted varieties of corn were resistant to this race. Therefore, corn breeders could not have predicted the need to incorporate race-T resistance into their new corn lines.

The second factor to consider is the extensive use of hybrid corn containing the T_{cms} genetic background. In the 1930s, breeders began to capitalize on the phenomenon of hybrid vigor. When two inbred lines are crossed or hybridized, the resulting seed corn will produce a crop with enhanced agronomic traits, including enhanced yield. To accomplish these crosses efficiently in corn, breeders must remove the male flowers from the female plant to prevent self-pollination. This was classically done by hand or machine. However, in the 1950s, cytoplasmic male sterility was discovered and incorporated in corn breeding programs. By 1965, nearly 80 percent of the entire U.S. corn crop was produced with male-sterile techniques, specifically the use of T_{cms}.

What the breeders did not know, however, was that hybrid corn with a T_{cms} genetic background was very susceptible to race T of *H. maydis*. In 1970, with proper weather conditions for disease development, with 85 percent of the corn crop containing T_{cms}, and with an abundant supply of race T inoculum, a southern corn leaf blight epidemic developed. Fortunately, however, the genetic basis for race-T susceptibility was quickly determined. By the next growing season, enough non-T_{cms} seed was available to farmers that losses were minimized.

Evidence for the molecular basis of T_{cms} activity has been obtained. Forde and Leaver (1989) reported that a polypeptide of 13,000 relative molecular mass (M_r) was unique to T_{cms} mitochondria and that its expression depended on the activity of a nuclear restorer gene (a gene that overcomes the effect of cytoplasmic sterility). Dewey et al. (1987) identified the mitochondrial gene encoding the 13,000 M_r polypeptide and determined that the protein was associated with the mitochondrial membrane. Rottmann et al. (1987) demonstrated that male sterile T_{cms} plants that mutated to male fertile plants lost their ability to produce the 13,000 M_r polypeptide and that the mutation occurred in the area of the mitochondrial genome that contains the gene for the 13,000 M_r polypeptide. In an effort to determine whether this polypeptide was also connected to increased susceptibility to *H. maydis*, Dewey et al. (1988) transferred the gene to *Escherichia coli* and demonstrated that bacteria producing this polypeptide were sensitive to *H. maydis* toxin. Therefore, the gene for the 13,000 M_r polypeptide may have a pleiotropic effect in that it confers both male sterility and susceptibility to *H. maydis*.

The story of T_{cms} is given here to illustrate the types of potential problems that have developed as a result of the introduction of new variants. The southern corn leaf blight epidemic was a highly publicized event: an epidemic ensued, and economic loss resulted. The year 1970 was certainly a bad year for corn production, but it was by no means a national catastrophe; corn production was back to almost normal within a year. Because an occasional unexpected crop loss may occur, it is important to have an arsenal of genetic modification techniques and genetic resources available that can be used promptly to limit unacceptable losses. New molecular methods for gene introduction will be beneficial in this regard.

Steady progress in the refinement of corn tissue culture systems (Vasil, 1988), coupled with the development of electroporation (Fromm et al., 1986) and particle-gun technologies (Klein et al., 1988), suggest that successful corn transformation may be imminent. Transgenic corn plants have been produced (Rhodes et al., 1988); although these plants were sterile, this accomplishment demonstrates that significant progress is being made to develop gene transfer systems for this important crop.

PAST EXPERIENCE WITH CONFINEMENT

Confinement is defined as any system of growing plants in which contact with plants of the same type is minimized or plants are kept in defined areas. Plant breeders traditionally use confinement procedures to minimize genetic contamination of their field plots by pollen from outside sources such as neighboring fields. In addition, confinement practices are used to keep plant pathogens from spreading into or out of experimental field plots. Agricultural research, therefore, has a long history of experimentation that has been confined or kept within bounds.

Both the private and the public sector, notably the land-grant institutions or the Agricultural Research Service of the United States Department of Agriculture (USDA), undertake the first of several stages of cultivar development. For example, cultivated varieties of wheat are the result of 7 to 14 years of research and testing by both the public and private sector before marketing (Table 3-1). During this time, small numbers of plants are grown at selected sites and kept under close observation for environmental or organismal effects on the plant. Extensive records are usually compiled and, in the public sector, summarized and published. Few lines (or potential varieties) survive such rigorous testing. Even after commercial use in farmers' fields, the plant's performance is examined periodically, by both sellers and producers of the seed or other propagative material. Some extremely well-adapted and highly productive cultivars have a long commercial life, because of desirable characteristics that are difficult to improve. Other cultivars survive only a short time, perhaps five years, before they are replaced by higher yielding, disease-resistant, or otherwise improved cultivars. Biotechnology has

TABLE 3-1 An Illustrative Wheat Breeding Program

Year	Generation	Activity	Area (acres)
1		Make 300 to 400 crosses between varieties or germplasm materials.	0.1
2	F_1	Grow in field, greenhouses, or both	0.1
3	F_2	Grow as bulk hybrid, evaluate for agronomic and disease traits; quality evaluations for milling, mixing curves, and protein content	0.5
4	F_3	Bulk seed select determined number of heads from best crosses	1.0
5	F_4	Head row nursery; 50,000 to 60,000 entries, screen for disease resistance, select 5% on basis of resistance and plant type	4.0
6	F_5	Preliminary observation nursery; agronomic value; disease resistance; quality evaluations for milling, mixing curves, and protein content	2.0
7	F_6[a]	Duplicate plots at one or more locations	2.3
8	F_7[a]	Preliminary yield trials at several locations	1.5
9	F_8[a]	Intrastate yield nursery at several locations	1.75
10	F_9[a]	Station plots, on-farm tests, regional nurseries, increase seed	4.0
11	F_{10-13}[a]	Repeat testing; large-scale milling and baking evaluations; seed increase; name and release to certified growers	30.0

[a]Quality evaluations for milling, baking, mixing properties, and protein content.

the promise to shorten the cycle to commercial availability by two or more years through specific gene transformation and identification of the particular genes conveying desirable attributes.

In the multiyear process of development of a useful cultivar, it is crucial to confine the seed and plants to the appropriate sites and to maintain the identity (purity) of the material (Table 3-2). This is done by confinement practices, which limit the plants or their products to a particular site and also protect neighboring fields from contaminating pollen. In this way, any unexpected effects can be observed. The distances cited in Table 3-2 are not absolute, but allow for acceptable levels of contamination. Specific information about the environment in which a cultivar was developed is necessary to make helpful site recommendations about suitable cultivars.

Confinement as practiced by plant breeders or plant pathologists may be achieved in several ways. Simple confinement may be accomplished by the choice of an isolated location. Border rows for

TABLE 3-2 Isolation Requirements for Production of Genetically Pure Seed for Certain Species of Field and Vegetable Crops

Type of Pollination	Species	Isolation Distance for Highest Level of Genetic Purity
Self-pollinated	Barley, oats, wheat, rice, soybean, lespedeza, field pea, garden bean, cowpea, flax grasses (self-pollinated and apomictic species)	Fields should be separated by a definite boundary adequate to prevent mechanical mixture 60 feet
Self-pollinated but to a lesser degree than those listed above	Cotton (upland type)	100 feet from cultivars that differ markedly
	Cotton (Egyptian type)	1320 feet
	Pepper	200 feet
	Tomato	200 feet
	Tobacco	150 feet or by four border rows of each cultivar. Isolation between cultivars of different types should be 1320 feet
Cross-pollinated by insects	Alfalfa, birdsfoot trefoil, red clover, white clover, sweet clover	600 feet 900 feet
	Millet	1320 feet
	Onion	5280 feet
	Watermelon	2640 feet
Cross-pollinated by wind	Hybrid field corn	660 feet (may be reduced if field is surrounded by specified numbers of border rows and the cultivars nearby are of same color and texture)
Grasses	900 feet	

ADAPTED FROM: Association of Official Seed Certification Agencies, 1971.

plants will limit both entry and exit of insects or diseases that might otherwise harm the plants of interest. Fencing limits animal access. In tests conducted on a small scale, one uses the smallest numbers of plants that will give the information desired. More elaborate barriers to limit dispersal beyond the site include removing pollinating organs from plants, bagging flowers, and adjusting the time of year the plants are grown to avoid insect pests. Multiple physical and biological barriers are used in research plots and often in commercial agriculture as well. Such barriers also include dams, soil terraces,

TABLE 3-3 Time Frames and Methods for Mitigating Unwanted Effects of Plants

Immediate (Hours to several days)	Short-term (0 to 3 years)	Long-term (More than 3 years)
Burning (eradication)	Breeding for resistance[a]	Breeding for resistance
Quarantine	Biological control[b]	Biological control
Tillage	Quarantine	Crop rotation
Chemicals[c]	Chemicals	Cultivar rotation
Biological control	Crop rotation	Soil amendments
Irrigation/flooding	Cultivar rotation	Weed control
Insect vector control	Irrigation/flooding	Erosion control
Machinery sanitation	Heat treatment	
Runoff water control	Soil solarization	
Solarization (cover with plastic)	Induced resistance	
	Meristem/tissue culture	
	Insect vector control	
	Weed control	
	Erosion control	

[a]Germplasm may be adequately identified for rapid development; otherwise the process normally takes 5 to 10 years.

[b]Few biological control agents are yet available for widespread use; several are under investigation and development for some disease-causing microorganisms.

[c]Choice and availability of chemical for target plant and associated microorganisms dictate feasibility and approach.

ADAPTED FROM: A. K. Vidaver and G. Stotzky, 1989.

tillage practices, and the use of chemical or biological agents for control of insects or fungi. If necessary, physical barriers and security against unauthorized persons may be needed.

Biological barriers include genetic modifications to produce sterility or to reduce the ability of the plant to survive or escape predators. The removal of reproductive organs and the removal of organisms that are hosts for a pathogen or insect can also be used. Death (normal decay), plowing under, and incineration are possible.

Collectively, these procedures work well in research and usually very well in commercial use to protect human health and the environment.

If these common practices lose effectiveness, various ways of mitigating deleterious effects are available (Table 3-3). Some of these means are inexpensive and can be applied quickly, while others may be costly and require longer periods to be effective. All these methods are applicable to genetically modified plants.

SUMMARY POINTS

1. Techniques of genetic modification of plants were divided into three broad categories for the purposes of this report: classical, cellular, and molecular. These techniques offer a wide array of possible genetic modification. Classical techniques include breeding by sexual hybridization, embryo rescue, undirected mutagenesis, and anther and ovule culture. Cellular techniques include cell fusion and somaclonal variation produced by tissue culture. Molecular techniques include directly introducing genes by a variety of transformation procedures.

2. The results of genetic modification of plants are usually divided into two categories: those that increase yield and those that increase reliability of performance. Although these modifications can affect the persistence of plants, it will be difficult to increase overall persistence of domesticated crops because many persistence-related traits have been eliminated through breeding.

3. Plant breeders have a long history of safe field testing and introduction of many genetically modified crops. When problems occur they have been manageable and for the most part confined to the managed ecosystem.

4. Routinely used methods of plant confinement offer a variety of options for limiting both gene transfer by pollen and direct escape of the genetically modified plant. Methods of confinement include biological, chemical, physical, geographical, environmental, and temporal control as well as limitation of the size of the field plot.

4
Enhanced Weediness: A Major Environmental Issue

GENERAL PRINCIPLES

Perhaps the single most commonly voiced concern about the introduction of genetically modified plants is that it might have the potential to inadvertently produce a new weed or increase the aggressiveness of existing weeds (R. K. Colwell et al., 1985; Tiedje et al., 1989).

This chapter discusses three aspects of the concern: whether the experience with the introduction of exotic plants into new environments (sometimes with the result that a weed problem is created) is a valid analogy for the introduction of genetically modified plants; the potential for domesticated crops to revert to a wild or weedy state; and the potential for hybridization between domesticated crops and wild relatives that might create or enhance weediness.

Evaluation of these issues first requires a careful definition of terms. The term "weed" has been variously defined, depending on the different perspectives of ecologists, agronomists, and the public. In this report we define a weed as an unwanted or undesirable plant in some human environments, that is, a plant that persists in human environments but is neither a crop (used for food, fiber, fuel, pharmaceuticals, or turf) nor an ornamental plant.

A characteristic of human environments and consequently a strong agent of selection among weeds is frequent disturbance, as occurs in arable fields, roadsides, foot paths, and the margins of reservoirs. Consequently, many plants that have become persistent weeds are species that arose earlier because their phenotypes permitted them to colonize special natural environments that exist in frequently disturbed sites. Such plants often display rapid growth, a short life cycle, high seed production, and long-distance dispersal of seeds (Baker, 1974). Not all colonizers are weeds, however, nor are all weeds colonizers.

Some weed species have also apparently required additional characters in order to thrive in close association with humans. These advantages include escape from biotic control agents such as predators, pathogens, and competitors (Harper, 1965). Such an escape is effective if a plant is suddenly transported far beyond its native range and therefore the range of one or more of its enemies. It is not surprising, then, that in most parts of the world, including the United States, the bulk of the weed flora are exotic plants (Holm et al., 1977; Smith, 1985; Mack, 1986), members of a species that enters a range in which that species has not occurred before (Mack, 1985). Perhaps most successful (most widespread, persistent, and abundant) are those weeds that have not only immigrated, but also have a long history of close association with human settlement (Baker, 1974).

Whether a plant becomes a weed depends on the relationship of the plant to its environment, especially with respect to control mechanisms that hold the organism in balance with that environment. A plant can become a weed if it escapes control by migrating to a new environment that lacks the factors that controlled the plant in its original habitat. In addition, a plant remaining in its original habitat may effectively escape a particular control factor, such as predation by a specific insect pest, by gaining a trait that imparts to it the ability to overcome the control factor. Any added trait that enhances performance (such as frost resistance or drought tolerance) would also be analogous to providing the plant with an advantage sometimes gained by plants in a new environmental range. Although this description is theoretically valid, it is necessary to keep in mind that there is extensive experience in these kinds of modifications in classical breeding. So far, weediness has not resulted from the addition of the traits of pest or herbicide resistance, nor frost or drought tolerance.

THE RELATIONSHIP BETWEEN THE INTRODUCTION OF EXOTIC PLANTS AND GENETICALLY MODIFIED PLANTS

The term exotic species, as used here, refers not only to entirely novel species in a new habitat, such as the Asian weed kudzu (*Pueraria lobata*) in the United States, but also to any species with an expanded geographic range, even when closely related plants are already present. In addition, exotic species usually refer to plants whose ranges were extended as a result of human intervention.

Ecological Implications of Introducing Plants with Many New Traits

Exotic species may not be strictly analogous to genetically modified organisms because many exotic species differ by many traits from any of their neighbors in the new environment. Consequently, the immigrants (such as *Agropyron repens*, *Eicchornia crassipes*, *Schinus terebinthifolius*) will owe their success in spread and eventual naturalization to a suite of characters (Holm et al., 1977; Barrett and Richardson, 1986; Morton, 1978). Genetically modified plants that are likely to be introduced in the near future (say, over the next 10 years) will differ by only one or a few traits from cultivated forms already in the same environment (the introduction of glyphosate-tolerant tomatoes).

Kudzu is a familiar example of a deliberately introduced exotic organism that has proven to have undesirable features. It illustrates the public's worst perceptions of errant organisms and simultaneously exemplifies an exotic organism that is not analogous to any hypothetical genetically modified organism. Originally introduced into the United States from China and Japan in the late nineteenth century for ornamental purposes, kudzu was eventually touted as an excellent stabilizer of soil embankments and as a forage crop on unproductive land. Cash incentives were even provided at one time to encourage farmers to plant it on abandoned fields (Miller, 1983).

By the 1950s, however, detrimental aspects of kudzu were recognized, as the vine often grows far beyond the site of its local introduction. It now commonly grows over forest canopies and telephone lines in the southeastern United States. Kudzu's success is based on a combination of features: it readily propagates vegetatively, it can grow on infertile soil and low amounts of soil water (Forseth and Teramura, 1987), and it has few (if any) serious parasites or herbivores in its new environmental range.

Kudzu exemplifies how the combined action of many traits introduced into a new environment results in a weediness problem. Our knowledge of invasions and particularly the characteristics that spell success or failure for immigrants is limited (Harper, 1982; Simberloff, 1985), despite the attempts to identify the putative characteristics of successful weeds (Baker, 1986; Bazzaz, 1986).

Ecologically Important Changes that Result from Small Genetic Alterations

Even though exotic species such as kudzu are not strongly analogous to genetically modified plants, circumstantial evidence suggests that a change in only a few characters can sometimes make a plant a successful invader. Within the large grass genus *Bromus* are several annual species that have become successfully naturalized in different temperate regions. *Bromus tectorum* spread rapidly in the interior Pacific Northwest in the early part of this century, whereas other members of the genus such as *B. mollis* and *B. brizaeformis* are much less common even though they were introduced earlier (Mack, 1981). In contrast, *B. mollis* is much more prominent than *B. tectorum* in the Central Valley of California, and *B. secalinus* can be a serious weed of cereals in northern Europe (Salisbury, 1961). The differences among these closely related species that explain their various success in new environmental ranges may be related to different tolerances to frost (*B. rigidus* and *B. rubens* are less tolerant than *B. tectorum*) and different flowering times (*B. japonicus* flowers before the onset of drought) (Hulbert, 1955). These species are morphologically similar and also share many ecologically important traits, yet they differ in their degree of success in their new ranges.

The exotic woody genus *Casuarina* provides another example in Florida. The two species, *C. equisetifolia* and *C. glauca* were deliberately introduced into southern Florida. The first has become a serious pest, while the second persists only locally. The most apparent difference between these closely related species is the inability of *C. glauca* to produce seed in the new range (Morton, 1980), which thereby limits its dispersal.

Other examples of environmentally important single-trait changes are demonstrated by the spread of *Chondrilla juncea* in

Australia and the role of insect herbivory in influencing the competitive ability of barley. *Chondrilla juncea* (skeletonweed) is a serious weed in the wheat-growing regions of southeast Australia. It has three morphological forms in Australia—termed A, B, and C—that differ most obviously in leaf shape, flower morphology, and fruit characteristics. Before a biological control program was initiated in the early 1970s, form A was much more widespread than the other two. But form A has proven to be much more susceptible than forms B and C to the deliberately released rust fungus, *Puccinia chondrillina*. As populations of form A have become infected, they have become less competitive than they had been, and their range has declined. Much of the range vacated by form A has been filled concomitantly with forms B and C (Burdon et al., 1981).

A similar reversal of competitive roles has also been documented between two cultivars of barley (*Hordeum vulgare*) that display a difference in their resistance to the aphid *Schizaphis graminum*. Under greenhouse conditions the aphid-resistant cultivar competes less well in mixtures of the cultivars. If the aphid is introduced into the mixtures, the competitive advantage of the susceptible cultivar is lost (Windle and Franz, 1979a; Windle and Franz, 1979b).

These examples illustrate that small genetic differences between closely related plants can produce phenotypes with different ecological properties that can increase or alter a plant's geographic range or enhance its aggressiveness in its normal range. How likely is this phenomenon for genetically modified crops or other plants being considered for field testing? Although most ecologically important traits remain unchanged, the interaction among these traits determines whether a species will become naturalized in a region. For example, a species could spread because it tolerates herbivores and parasites and tolerates some aspects of the physical environment (such as salinity) in the new range. Gottlieb (1984) compiled a list of diverse traits in plants that can be governed by one or a few genes. Whether the plant is erect or prostrate, branched or not, an annual or a biennial, or bears its leaves basally or higher on the stem can all be governed by a few genes. The suggestion from this list of traits is that major changes in plant architecture and subsequent performance could be achieved through rather small gene changes or insertions by recombinant techniques. Such changes in architecture would be readily detectable in greenhouse and field tests. The likelihood that these changes would occur randomly (and be retained) is very small.

We do not know to what extent successful naturalization of exotic organisms hinges on their possession of one or a few traits rather than a group of characters. Multiple genes inducing multiple traits should increase the probability of assembling an organism that can cause ecological changes if grown on a large scale.

Several exotic species (for example, *Cytisus scoparius*, *Ulex europaea*, *Leucaena leucocephala*) owe much of their successful naturalization to their ability to fix nitrogen in a new environment that is chronically low in nitrogen (Vitousek, 1986). Nitrogen fixers, such as *Alnus* spp., characteristically are the first invaders on newly formed volcanic soils. *Myrica faya*, a small exotic tree, is rapidly altering the nitrogen balance on volcanic sites in the Hawaiian islands. As the nitrogen content of the volcanic soil has increased, new species have become established on these sites (Vitousek, 1986).

Relatively minor genetic changes can produce plants with altered ecological properties, a phenomenon plant breeders have capitalized on for decades; for example, introducing a single gene in wheat can impart resistance to a specific race of stem rust. Similarly, herbicide-resistant canola and soybean plants have been produced by minor genetic changes. Such changes have not resulted in increased weediness of these widely used crops.

THE ABILITY OF CROPS TO REVERT TO A WILD OR WEEDY CONDITION

Crops that have been subjected to long-term breeding (for example, beans, maize, and wheat) are less likely to revert to a wild state than crops that retain many wild characters (artichokes, forage grasses, and grain amaranths). Highly domesticated crops have lost their ability to compete effectively with the wild species in natural environments. Domesticity arises because many characters that would enhance weediness (seed shattering, thorns, seed dormancy, and bitterness) have been deliberately eliminated from the crop plant through intensive breeding efforts. The reassuring history for cultivated crops does not completely preclude a genetically modified crop from becoming weedy, but it suggests that the likelihood of that event is small. As new traits are inserted into cultivated crops, they might possibly change the crop in an ecologically significant way, but past experience with classical breeding has shown this to be a manageable problem. Field trials should identify such possibilities.

The descendants of crops may become weeds in agricultural

fields, and in some circumstances they may move beyond the boundaries of the field and become weeds in seminatural or even natural communities. More than a decade ago, Harlan (1975) compiled an often-cited list of the wild races of crops that included many row crops such as beets, cabbages, and watermelons. The relevance of these examples depends, in part, on the level of domestication in the crop.

Some crops such as artichoke, sugar beets, and some citrus (Gade, 1976; Pickersgill, 1981; Thomsen et al., 1986), seem prone to become weedy. The ability to revert to a weedy condition has never been attributable to traits deliberately retained in the domesticated crops—that is, traits that have been the object of an active breeding program.

HYBRIDIZATION BETWEEN CROPS AND THEIR WILD RELATIVES

Two closely related ecological questions that may be important to the introduction of genetically modified plants are (1) Does hybridization between crops and their wild relatives result in transfer of traits from the cultivated form to the wild relative? and (2) Does such gene flow increase the weediness of wild relatives? If the opportunity exists for the transfer of genetic traits from a genetically modified organism to a wild (and potentially weedy) relative, a potential problem exists. The problem poses three relevant questions: (1) Does the genetically modified crop have extant relatives? (2) What is the extent of hybridization between crop and relatives in nature? and (3) What is the current ecological role of the relative in natural ecosystems?

Practically all crops have wild relatives at some taxonomic level. The more important question is whether wild relatives occur in the range in which the genetically modified crop is grown or will be grown. The answer varies, as no one region of the world includes the home range of most crops, although arid central Asia and Asia Minor are the centers of origin for many crops (Table 4-1). Southeast Asia includes the home range of many weeds. Temperate North America, especially the United States, includes the home ranges for very few crops, as U.S. agriculture is based largely on crops of foreign origin. This paucity of crops derived from North American sources means there will be relatively few opportunities for hybridization between crops and wild relatives in the United States, except where both

TABLE 4-1 Crops and Their Probable Regions of Origin. (Note That Comparatively Few Crops Are Native to North America)

Crop	Scientific Name	Common Name
	EUROPE AND TEMPERATE ASIA	
Cereals	Avena sativa L.	Oats
	A. strigosa Schreb.	Fodder oats
	Hordeum vulgare L.	Barley
	Secale cereale L.	Rye
	Triticum aestivum L.	Bread wheat
Pulses	Cicer arietinum L.	Chick-pea
	Lens esculenta Moench	Lentil
	Pisum sativum L.	Garden pea
	Vicia faba L.	Broadbean
Root and tuber crops	Beta vulgaris L.	Beet, mangel, chard
	Brassica rapa L.	Turnip
	Daucus carota L.	Carrot
	Raphanus sativus L.	Radish
Oil crops	Brassica campestris L.	Rapeseed
	Carthamus tinctorius L.	Safflower
	Linum usitatissimum L.	Flax, linseed
	Olea europea L.	Oli
Fruit and nuts	Ficus carica L.	Fig
	Juglans regia L.	English walnut
	Phoenix dactylifera L.	Date palm
	Prunus amygdalus Stokes	Almond
	P. armeniaca L.	Apricot
	P. avium L.	Cherry
	P. domestica L.	Plum
	Pyrus communis L.	Pear
Vegetables and spices	Cucumis melo L.	Melon
	Allium cepa L.	Onion
	A. sativum L.	Garlic
	Brassica oleracea L.	Cabbage, cauliflower, brussels sprouts, kale, kohlrabi, broccoli
	Cucumis sativus L.	Cucumber
	Lactuca sativa L.	Lettuce
Forage crops	Bromus inermis Leyss.	Smooth bromegrass
	Dactylis glomerata L.	Orchardgrass, cocksfoot
	Festuca arundinacea Schreb.	Tall fescue
	Medicago sativa L.	Alfalfa
	Phleum pratense L.	Timothy
	Trifolium spp.	The true clovers
Drug crops	Digitalis purpurea L.	Digitalis
	Papaver somniferum L.	Codeine, morphine, opium

Table 4-1 (continued)

Crop	Scientific Name	Common Name
	AFRICA	
Cereals	Oryza glaberrima Steud.	African rice
	Pennisetum americanum (L.) K. Schum.	Pearl millet
	Sorghum bicolor (L.) Moench	Sorghum
Pulses	Vigna unguiculata (L.) Walp.	Cowpea
Root and tuber crops	Dioscorea cayenensis Lam.	Yam
Oil Crops	Elaeis guineensis Jacq.	Oil palm
	Ricinus communis L.	Castor oil
Fruits and nuts	Colocynthis citrullus (L.)	Watermelon
Fiber plants	Gossypium herbaceum L.	Old world cotton
Forage crops	Cynodon dactylon (L.) Pers.	Bermuda grass
	Digitaria decumbens Stent	Pangolagrass
	Eragrostis lehmanniana	Lovegrass
	Panicum maximum Jacq.	Guineagrass
Drug plants	Coffea arabica L.	Coffee
	CHINA	
Cereals and pseudocereals	Fagopyrum esculentum Moench	Buckwheat
	Oryza sativa L.	Rice
	Panicum miliaceum L.	Proso millet, broomcorn millet
	Setaria italica (L.) Beauv.	Italian millet, foxtail millet
Pulses	Glycine max (L.) Merr.	Soybean
Root and tuber crops	Brassica rapa L.	Turnip
	Dioscorea esculenta (Lour.)	Chinese yam
Oil Crops	Brassica campestris L.	Rapeseed
	B. juncea (L.) Czern. & Coss.	Mustard seed oil
Vegetables and spices	Alium bakeri Regel	Chinese shallot
	Cinnamomum cassia Blume	Spice
	Cucumis sativus L.	Cucumber
	Zingiber officinale Roscoe	Ginger
Drug plants	Camellia sinensis (L.) Ktze.	Tea
	Cinnamomum camphor[a] (L.)	Camphor tree

Table 4-1 (continued)

Crop	Scientific Name	Common Name
	SOUTHEAST ASIA AND PACIFIC ISLANDS	
Cereals and pseudocereals	Oryza sativa L.	Rice
Oil crops	Cocos nucifera L.	Coconut
	Sesamum indicum L.	Sesame
Fruits and nuts	Citrus aurantiifolia Swingle	Lime
	C. aurantium L.	Sour orange
	C. limon (L.) Burm. f.	Lemon
	C. nobilis Lour.	Tangerine
	C. paradisi Macfad.	Grapefruit
	C. sinensis (L.) Osb.	Sweet orange
	Musa acuminta Colla	Banana (A genome)
	M. balbisiana Colla	Plantain (B genome)
Vegetables and spices	Elettaria cardamomum (L.) Maton	Cardamom
	Syzygium aromaticum (L.) Merr. & Perry	Clove
	Myristica fragrans	Nutmeg
	Piper nigrum L.	Black pepper
	Solanum melongena L.	Eggplant
Starch and sugar plants (not roots)	Saccharum officinarum L.	Sugarcane
	MESOAMERICA AND SOUTH AMERICA	
Cereals	Zea mays L.	Corn
Fruits and nuts	Anacardium occidentale L.	Cashew
	Ananas comosus (L.) Merr.	Pineapple
	Bertholletia excelsa HBK.	Brazil nut
	Carica papaya L.	Papaya
	Carica candicans A.	Gray papaya
	Persea americana Mill.	Avocado
	Psidium guajava L.	Guava
Vegetables and spices	Capsicum annuum L.	Pepper
	Capsicum baccatum L.	Pepper
	Cucurbita maxima L.	Squash
	Cucurbita pepo L.	Squash, pumpkin
	Phaseolus vulgaris L.	Bean
	Lycopersicon esculentum Mill.	Tomato
	Solanum tuberosum L.	Potato
	Vanilla planifolia Andr.	Vanilla

Table 4-1 (continued)

Crop	Scientific Name	Common Name
Fiber plants	Gossypium hirsutum L.	Upland cotton
Drug plants	Nicotiana tabacum L.	Tobacco
	Theobroma cacao L.	Cacao, chocolate
	NORTH AMERICA	
Oil crops	Helianthus annus L.	Sunflower
Fruits and nuts	Vitis labrusca L.	Fox grape
	V. rotundifolia Michaux.	Muscadine grape
	Vaccinium macrocarpon Aiton	Cranberry
	Vaccinium (several species)	Blueberry
	Fragaria (several species)	Strawberry
	Rubus idaeus Richardson	Red raspberry
	Rubus (several species)	Blackberry
	Rubus (several species)	Dewberry
Vegetables	Helianthus tuberosus L.	Jerusalem artichoke

ADAPTED FROM: Harlan, 1975.

crop and wild relatives have immigrated (Table 4-2). The incidence of hybridization between genetically modified crops and wild relatives can be expected to be lower here than in Asia Minor, southeast Asia, the Indian subcontinent, and South America, and greater care may be needed in the introduction of genetically modified crops in those regions.

If a crop has no relatives within the distance its pollen can travel, no hybrids will develop. Spatial separation is an obvious barrier to hybridization, but only anecdotal knowledge exists on the actual limits of pollen transport (Ellstrand, 1988). Furthermore, even if relatives are nearby, there is no assurance that viable hybrids will be produced, as there often are many formidable barriers to gene flow, such as differences in ploidy level, flowering time, and breeding systems (Simmonds, 1979). In fact, the deliberate introduction of genes from wild relatives into certain crop species by classical breeding techniques has been achieved only by manipulating the flowering time and by repeated hand pollination (as in potatoes). Even if fertilization is accomplished naturally, there is no assurance that

TABLE 4-2 Some Crops Growing Sympatrically in the United States with Congeners or Wild Races with Which Natural Hybridization Is Possible.

Crop	Primary Gene Pool
Sorghum bicolor (sorghum)	S. halepense (Johnson grass)
Raphanus sativus (radish)	R. raphanistrum (wild radish)
Setaria italica (foxtail millet)	S. italica frequently naturalized as a weed, may not exist in the United States
Brassica rapa (turnip)	B. campestris (wild type)
Brassica campestris (rape)	B. campestris (wild type, field mustard)
Amaranthus cruentus (amaranth); A. caudatus; A. hypochondriacus	A. hybridus; A. powellii; A. retroflexus
Beta vulgaris (beet)	May have only weedy race in Europe
Daucus carota (carrot)	D. carota spp. carota
Helianthus annuus (sunflower)	H. annuus (wild morphs); H. bolanderi
Cucurbita pepo (squash, pumpkin)	C. texana (wild marrow)
Secale cereale (rye)	S. cereale and S. montanum
Lactuca sativa (lettuce)	L. serriola
Avena sativa (oat)	A. fatua (wild oat)
Cynara scolymus (artichoke)	C. scolymus (wild types)

ADAPTED PRIMARILY FROM: N. W. Simmonds, ed., 1979, and references therein.

further plants will be produced. For a gene to pass between relative and crop and be permanently incorporated into either the crop or the relative, introgression (introduction of a gene from one gene complex into another) must occur regularly (Anderson, 1949). This occurs at exceedingly low frequency in many crops and wild relative combinations.

Evidence for gene introgression by hybridization between crops and wild relatives has often been only circumstantial. Because plant breeders are usually concerned with the detection and elimination of wild traits in a crop, the low incidence of documented transfer and introgression that occurs from crops to wild relatives may be an artifact. A complication in reliably identifying such introgression continues to be the possibility of convergent evolution between crop and wild relatives. The mechanism by which this could occur is easily envisioned: An agronomic practice such as seed sorting by size imposes strong directional selection in a wild relative (or even an unrelated weed) for those phenotypes with the same seed size as the crop (Barrett, 1983). Seed size, shape, and even color can be remarkably similar between the crop and the weed without hybridizations occurring.

Forty years ago, plant breeders in India selected for increased anthocyanin production in cultivated rice in an attempt to improve the ability of paddy workers to discriminate between otherwise indistinguishable seedlings of cultivated and wild rice (*Oryza* species). Although the cultivated rice seedlings were readily identified at first by their purple leaves, within several plant generations the trait had been transferred to the wild relative, thus rendering the trait useless from a cultivation standpoint (Parker and Dean, 1976). Other putative examples of gene flow from crop to wild relative have been reported for crosses between corn and teosinte, "Eastern Carrot" and wild carrots, "kayseri" alfalfa and weedy relatives, and between durum wheat and wild emmer wheat. The evidence is mainly morphological and therefore subject to alternative interpretations (for example, convergence after mutation in the wild relative and subsequent directional selection) (Small, 1984).

Other examples, also based largely on morphological evidence, occur among the cultivated *Amaranthus caudatus*, *A. cruentus*, and *A. hypochondriacus*. Each of these species forms hybrids with one or more weedy amaranths in California and Mexico. Gene flow to the weedy amaranths is probably more obvious and persistent because of the strong selection by hand-cultivation against the preservation of hybrids with the wild parent's trait of dark seed (Sauer, 1967; Tucker and Sauer, 1958). In the Sacramento-San Joaquin delta, Tucker and Sauer (1958) identified many amaranth hybrids that resulted from crosses between crop and wild relatives. They maintained, without direct evidence, that under cultivation in the light, highly fertile organic soil in the region, hybrids could out-compete their weed parents (*A. hybridus*, *A. powellii*, and *A. retroflexus*) because they had acquired traits from their crop parents for more robust stature and high fecundity.

Gene flow has apparently occurred from crop to wild relative in rye (*Secale* spp.) in California, where a weedy rye probably derived from a cross between *S. cereale* and *S. montanum* has become increasingly crop-like through introgression with the cultivated *S. cereale.* This introgression has proceeded to such an extent that farmers have abandoned efforts to grow cultivated rye for human consumption and are deliberately sowing the hybrids for forage (Jain, 1977; Suneson et al., 1969). In each of these examples, the putative transfer of a trait from the crop to the wild relative has resulted in the relatives' becoming more similar to the crop; in the above-cited example with Asian rice, the introgression resulted only in an enhancement of mimicry of the crop.

Evidence is restricted to morphological or cytological similarities between the crop and the wild relative. However, much of the evidence is circumstantial rather than experimental; clear demonstration of introgression depends on molecular analyses of isozymes or other techniques. Recent work with molecular marker loci has refuted several earlier claims of introgression in *Helianthus* (Rieseberg et al., 1988) and some reports of introgression between maize and teosinte (Doebley, 1984). Even with isozyme studies there is the possibility for an alternative interpretation; the crop and the wild relative may share alleles derived from a common ancestor rather than through more recent introgression. Consequently, the best evidence for recent gene transfer arises in cases in which a wild relative possesses alleles in common with a crop, but only in those populations that have recently come into sexual contact with the crop.

Convincing evidence for a transfer of genes from a crop to a wild relative does exist in several crop-weed complexes: African rice, maize, and *Cucurbita.* Second (1982) has shown that African rice, *Oryza breviligulata*, contains more isozymic variation than cultivated rice, *O. glaberrima.* His data suggest that this variable weedy rice arose through introgression between the wild form and cultivated rice. Doebley (1984) found evidence for introgression of cultivated maize into *Zea diploperennis*; one plant possessed two alleles that had not been found previously in the wild species but that are common in maize. Because the two loci are tightly linked, there is at least the strong suggestion that the chromosome segment carrying these loci was transferred through hybridization into *Zea diploperennis.* In the southern United States the cultivated *Cucurbita pepo* (squash) occurs in the same area as the wild species, *Cucurbita texana* (Texas gourd). Decker and Wilson (1987) found that alleles typical of the cultivar can occur in the wild species. This introgression enhanced weediness in the sense of making the hybrids more difficult to distinguish from the crop (that is, their mimicry of the crop increased), but the hybrid was no more aggressive, nor did it have an enhanced ecological range. Consequently, the products of the inadvertent transfer of crop genes to relatives have been confined to the field in which the plants were grown. From the standpoint of eradicating the weeds, the result of this introgression is at worst undesirable.

The hybridization between cultivated sorghum and one or more of its wild relatives is more serious. "Hybrid grain sorghum" (*Sorghum bicolor)* is produced through the cytoplasmic male-sterility method in which two inbred lines are hybridized. The seed is harvested from the male-sterile plant. If pollen of one or more weedy sorghums is inadvertently allowed to fertilize the stigmas of the male-sterile plants, the offspring are useless commercially and represent a genetically diverse cluster of races and "off-types" called shattercane. These plants usually express many traits of the wild parent, such as the perennial habit (inherited from *Sorghum halepense*), height, or self-sowing seed (Baker, 1972), a trait inherited from *Sorghum sudanense.*

Hybrids bearing traits of *S. halepense* (Johnson grass) present potentially serious weed problems because the vegetatively vigorous *S. halepense* is eradicated only with great difficulty and expense (Holm et al., 1977; Warwick and Black, 1983). The direct role of introgression with the cultivated sorghums in the enhancement of weediness in *S. halepense* is not clear, but the circumstantial evidence at least suggests the production of more persistent plants. De Wet (1966) maintains that *S. halepense* in its native range in the Old World has never been an excessively weedy plant and that its weediness was enhanced coincidentally with its introgression with cultivated sorghum in the United States. Acquisition of these traits is unusual in that their advantage to the weedy offspring is not confined to enhancing the weed's mimicry of the crop. If these Johnson grass populations extend their already major ecological role outside agricultural fields, they will represent the most extreme category of known risk associated with gene flow from crop to weedy relative. Biotypes of *S. halepense* in the northeastern United States apparently have acquired traits of ecological importance through introgression, including such crop-like features as earlier flowering, greater seed production, larger individual seed weight, and subsequently more rapidly emerging seedlings than other biotypes (Warwick et al., 1984).

The male sterility method produces a similar, although less serious, weed problem in the cultivation of sugar beets for seeds in northern Europe. If these male-sterile plants are inadvertently pollinated by the pollen of *Beta vulgaris* subsp. *maritima* (wild sea beet), some cultivar x wild F_1 hybrid seed eventually will be produced in the crop field (Pickersgill, 1981).

While hybridization between a crop and its wild relative may not be prevented by morphological, cytological, and developmental barriers, there is little likelihood that domesticated traits will be retained in a wild relative. Much of the emphasis in plant breeding has been toward traits that would reduce adaptation to the wild (for example, enhanced oil content in the seed, or an enlarged fleshy root), especially if enhanced production for these features came at the expense of plant fitness. Important commercial traits, such as pest resistance, that have the potential to alter the ecology of wild relatives have not been a problem with the possible exception of gene transfer from cultivated sorghum to Johnson grass.

SUMMARY POINTS

1. The analogy between the introduction of an exotic species into a new environment and the introduction of a genetically modified crop plant is tenuous because introduced exotic plants that have caused problems bring with them many traits that enhance weediness, whereas genetically modified plants are modified in only a few characteristics.

2. Genetic modifications of only a few genes can produce a modified plant with significant, ecologically important alterations. However, genetically modified crops are not known to have become weedy through the addition of traits such as herbicide and pest resistance.

3. Domesticated crops, such as soybeans, corn, and wheat, have been genetically modified to such an extent that they can no longer compete effectively with wild species in the natural ecosystem. These crops are unlikely to revert to a weedy condition upon further genetic modification. Some forage grasses are more likely to revert to a weedy condition.

4. Most crop plants in the United States are not native, and, unless weedy close relatives have been imported, no close relatives with which the crop might hybridize are present. However, where cross-hybridizing wild relatives do exist in close proximity (such as the sunflower), precautions may be necessary to limit gene flow from the crop to the wild relative. Gene introgression, when demonstrated, has often caused the wild species to become more like the crop, with consequences of enhanced weediness of the wild relative largely confined to agricultural fields.

5
Past Experience with the Introduction of Modified Plants: Molecular Genetic Techniques

Contemporary methods of genetic modification offer unique advantages for crop improvement. They complement existing plant-breeding efforts by increasing the diversity of genes and germplasm available for incorporation into crops. The directed transformation of commercial varieties and hybrids should significantly shorten the time to commercial release.

The rapid progress that has been made in gene identification and isolation methods, plant tissue culture, and gene transfer techniques has now permitted the extension of specific genetic change (for example, by recombinant DNA methods) to more than 30 species of crop plants (Gasser and Fraley, 1989). Today, nearly all major dicotyledonous crop species, including row crops (cotton, soybeans), vegetables (tomato, potato), forages (alfalfa, clover), and trees (poplar, pear), can be genetically modified by molecular methods. New methods have facilitated the development of transformation systems for use in corn and other monocotyledonous crops. Within the next few years, all major crops will probably be amenable to improvement through molecular approaches as a matter of course.

PROPERTIES OF MOLECULAR GENETIC MODIFICATIONS

Methods of Gene Introduction

A variety of techniques have been developed to introduce genes successfully into recipient plants. These techniques can be broadly grouped into those involving biological carriers, vectors, and those involving physical, or nonvectored, methods. Physical methods for modification include introducing DNA fragments into cells by microinjection (Crossway et al., 1986) or electroporation (Fromm et al., 1986) or introducing DNA bound to a metal microparticle that is accelerated into target cells (Klein et al., 1988; Johnston et al., 1988). Delivering genes by physical methods can produce a "simple" pattern of DNA insertion (a single DNA fragment inserted at one chromosomal location) or a "complex" pattern (multiple DNA insertions at one or more genetic loci).

One of the vectored methods commonly used for plant modification utilizes nature's own genetic engineer, *Agrobacterium tumefaciens* (Fraley et al., 1986; Bevan, 1984). By deleting the genes that modify normal cells into tumorigenic cells and leaving intact those genes that are responsible for transferring DNA from the bacterium to the plant cell nucleus, modified *A. tumefaciens* cells can vector desirable genes into appropriate plants cells. Between 70 and 80 percent of the transformed cells produced by this method have a single target gene inserted at a single locus (simple pattern). Other vectored methods for introducing genes include the use of plant DNA (Brisson et al., 1984) and RNA (French et al., 1986) viruses.

Gene insertion, whether transferred by vectored or nonvectored methods, appears to be a random event. Methods have not yet been developed to target the insertion of a gene to a specific chromosomal location, although progress is being made (Paszkowski et al., 1988). Because of the random nature of gene insertion, some transgenic cells or plants may exhibit more or less gene expression than others. Thus, it is usually necessary to evaluate several different transgenic plants to choose those with the desired levels of gene expression. This type of selection is reminiscent of the evaluation done by plant breeders as plant lines are developed.

In most modification methods, transgenic cells are selected in some manner (generally by resistance to an antibiotic or by screening with an appropriate gene reporter system) and exposed to altered cultural regimens to induce regeneration of plants. Regenerating plants from single cells, calli, or explants may produce somaclonally variant plants that, because of the cell culture or regeneration process, have a different phenotype than the parent. For example, somaclonally variant plants may be altered in terms of ploidy, sterility, or patterns of plant development. It is imperative that variants be recognized as such and not be confused with the direct products of modification per se.

Genetic Stability of the Alteration

Genetic modification usually involves the introduction of DNA into nuclear chromosomes and expression of the gene as a dominant trait. Such introduced genes, studied in a number of different settings, have been found to be inherited with stability equal to that of other nuclear genes. For example, Nelson et al. (1988) performed a limited field test with tomato plants that were stable as fourth-generation progeny after modification. Other workers have reported similar findings with genes introduced into other plants (Wallroth et al., 1986; Deroles and Gardner, 1988). No evidence suggests that introduced genes are lost more or less frequently than other plant nuclear genes. In addition, no evidence suggests that gene insertion with *Agrobacterium*-based vectors imparts any plant pest characteristics to the recipient plant.

The vast majority of plant modifications target nuclear chromosomes; however, attempts are being made to modify organellar genomes, those of chloroplasts and mitochondria. Whereas modification of *Chlamydomonas* chloroplasts has been demonstrated (Boynton et. al., 1988), similarly reproducible results with higher plant cell chloroplasts remain to be established. When these are achieved, the nature of inheritance of the introduced gene will be changed because chloroplasts are transmitted maternally, but usually not through pollen. Modification of organelles will probably be important for engineering such traits as herbicide resistance and male sterility.

There currently are a few replicon-based (autonomously replicating) vector systems for plants (Brisson et al., 1984; French et al., 1986; Grimsley et al., 1987); however, these have not achieved the same utility as in bacteria and yeast. It is likely that replicons based on RNA-containing or DNA-containing plant viruses will be developed to induce desired proteins or nucleic acids in plants. In some cases, part or all of the viral genome may be integrated into the nuclear genome and then regulated for expression in specific cell types. In other instances, a modified virus or part thereof may be introduced and expressed as a replicon per se. Because few plant viruses are efficiently seed-borne, such replicon-based systems will probably not be used widely for introducing agronomic genes into plants.

In all gene delivery and gene expression systems discussed (other than viral-based replicons), it is highly unlikely that the new gene will be transmitted to different plant types other than through sexual means. Thus, while *A. tumefaciens* modification involves the use of the modified bacterium to deliver the gene, the bacterium itself is removed after gene introduction is completed. This is accomplished by treating the transformed cells and regenerating plantlets with antibiotics that kill the bacterium. Collecting seed from transgenic plants excludes *A. tumefaciens*, which further ensures that the bacterium does not contaminate the progeny.

No evidence to date exists that stably integrated DNA is likely to be transferred by mechanisms other than hybridization under natural conditions, by either insects or microorganisms. Thus, there is no logical basis for more concern with the unusual transfer of an introduced nuclear gene than with any other nuclear gene transfer.

The types of genetic alterations that have been achieved to date include the transfer of large segments of DNA (a segment as long as 50 kilobases of DNA); an upper size limit for transfer has not been determined. Generally, much smaller segments of DNA, from less than 2 to 10 kilobases, are introduced. Gene transfers could theoretically include many genes, although practical considerations generally mitigate against transferring more than four or five genes at any one time. Multiple transformation of a single individual could produce a plant with many introduced genes, as does sexual hybridization of individuals that carry genes at distinct alleles.

Whereas genes are commonly introduced to add new traits, it has not been possible to inactivate or remove a specific gene by homologous recombination or insertional activation. However, an alternative approach that emphasizes antisense gene constructs has been successful in eliminating or reducing the expression of endogenous genes. Several applications of antisense (nucleic acid sequences that are complementary to sequences that code for a protein) technology in plant systems have been described, including the alteration of chalcone synthase genes (Van der Krol et al., 1988) and alterations to produce tomato fruit deficient in polygalacturonase that retain firmness for an extended period (Sheehy et al., 1988).

Types of Genetic Alterations

During the past 5 years a variety of genes have been introduced into plants for research purposes, but relatively few have the potential for use in agriculture and food production. Those of likely importance to production agriculture (Boyce Thompson Institute, 1987) in the near future include

- plants that express a gene that induces accumulation of insecticidal proteins, including *Bacillus thuringiensis* endotoxins (Fischhoff et al., 1987; Vaeck et al., 1987) and a variety of protease inhibitors (Hilder et al., 1987). Such proteins will limit the feeding of insects on the modified plants and reduce the need for chemical insecticides.
- plants that contain genes that encode the capsid protein of one or more plant viruses. The accumulation of viral capsid proteins protects these plants against the virus from which the gene was taken as well as against closely related viruses (Powell-Abel et al., 1986; Tumer et al., 1987).
- plants that are resistant to specific herbicides or classes of herbicides because they either detoxify the herbicide or resist its effects (Shah et al., 1986; Stalker et al., 1988; Haughn et al., 1988). The resistance traits will make it possible to use in agriculture normally nonselective but readily degraded herbicides that are safe to other life forms, thereby reducing weed control costs and long-lasting chemical damage to the environment.
- plants whose flower colors are altered (van der Krol et al., 1988), fruits remain firm (Sheehy et al., 1988), and seed protein or oil compositions are altered (Beachy et al., 1985; Sengupta-Gopalan et al., 1985).

The rapidly expanding knowledge base in plant biology makes it likely that future targets for plant improvement via molecular genetic techniques will include resistance to environmental pressures that can affect plant productivity. This could include resistance to heat, drought, flooding and salt stresses, pathogenic bacteria, fungi, and parasitic nematodes. In addition, these tools should significantly increase our current understanding of plant development and gene expression (Goldberg, 1988).

CASE STUDIES OF PLANTS MODIFIED BY MOLECULAR GENETIC TECHNIQUES

The field research on crops genetically modified has been less controversial than environmental introductions of other organisms. This may be attributed to the use of domesticated plants with which we have substantial experience regarding confinement during field research. During 1987-88, more than 20 trials were approved for field research with plants modified by molecular means including tomato (14 trials) and tobacco (7 trials) (Animal and Plant Health Inspection Service, unpublished, 1989). Requests up to March 1, 1989, include an increasing number of agronomic crops: cotton (3), soybean (3), alfalfa (2), potato (2), and rice (1) as well as additional tomato (4) and tobacco (1) trials. Of the 36 approved thus far and of those requested trials as of March 1, 1989, only one is from a noncommercial research group.

These requests and approvals are mainly for additions of single genes for resistance to herbicides (18), insects (19), and viruses (8), and a DNA sequence addition that enhances fruit quality (2). Results of these introductions have not raised any additional safety concerns.

The tests have taken place at diverse locations across the United States, including Illinois, Florida, California, Mississippi, Wisconsin, Delaware, and North Carolina. All were reviewed in detail by the Department of Agriculture with review and inputs by other governmental agencies. The key consideration in approval of each test has been a scientific evaluation of its risk and environmental impact. The major issues that have emerged from these discussions are

- stability of the inserted genes,
- undesirable alteration in crop phenotype,
- environmental impact on nontarget species,
- potential for weediness of genetically modified crops, and
- ability to maintain the gene within the test site.

Stability of Inserted Genes

Crop plants modified by molecular techniques have been produced either with *A. tumefaciens* Ti (for tumor inducing) plasmid vectors or by a variety of nonvector-mediated methods such as microinjection, electroporation, particle guns, or calcium-phosphate precipitation. Tens of thousands of plants in over 30 different crop species have been studied in contained facilities with respect to gene expression and inheritance patterns. The cumulative results demonstrate that the introduced DNA sequences are incorporated into random sites in the genome, stably maintained through both vegetative and reproductive propagation, and neither excised nor transferred.

All the evidence indicates that genes or traits introduced by molecular methods behave similarly to those introduced by classical techniques such as cell selection, mutagenesis, or sexual hybridization—that is, regular inheritance patterns in generations (Fraley et al., 1986; Kuhlemeier et al., 1987).

Undesirable Alteration of Plant Phenotype

Since gene insertion is random, inactivation of an important plant gene or genes could possibly result from the insertion process. the data accumulated to date, however, do not support this possibility. Efforts to introduce DNA to isolate genes by insertional inactivation reveal it to be an event of extremely low probability. The low frequency is understandable because less than 5 percent of the DNA in typical crop plant genomes constitutes actively expressed genes, and, in many cases, plant gene families may contain 5 to 10 functional members (Goldberg, 1988). Inactivation of a single gene, therefore, is unlikely to produce an altered phenotype. There has been a recent report, however, of gene inactivation by transferred DNA (T-DNA) insertion in *Arabidopsis thaliana* (Feldman et al., 1989). The *Arabidopsis* haploid genome, however, consists of only 70,000 kilobases, which is about 1/80 the size of the wheat genome. The small genome size of *Arabidopsis* greatly increases the likelihood that insertional mutagenesis will lead to gene inactivation.

Although the inserted gene or gene product might be able to impair some important plant process through an unknown mechanism, such a risk is no greater than that associated with classical breeding. With the molecular modifications, the introduced sequences and their functions are known precisely, and their functioning in a new genetic background can be experimentally determined in greenhouse studies and in small-scale field tests. A variety of molecular probes are available to monitor the location, expression, and function of introduced genes.

Recombination of DNA sequences is a normal consequence of sexual hybridization and an important contributor to the generation of new varieties and hybrids as shown by restriction fragment length polymorphisms (RFLPs) (Tanksley et al., 1989). The existing procedures for plant breeding, field evaluation, and crop certification have evolved to deal with the consequences of genetic recombination. Off-types displaying undesirable phenotypes are removed (rogued) as standard procedure. In a history of 75 years of breeding and crop testing, crop breeders have been successful in protecting against the introduction of undesirable traits into crop varieties; the earlier described southern corn leaf blight, by contrast, was one example of an undesirable phenotype that went undetected.

Environmental Impact on Nontarget Species

Some people are concerned that crops modified by molecular techniques may have an adverse impact on the environment. These issues involve managed and natural ecosystems (which are addressed in this report) and the possible risk to animal and human health (which is not considered here).

Risks to natural and managed ecosystems focus on the altered plants becoming weeds in succeeding crops or on the movement of genes to wild relatives that would increase the weediness of those relatives. These aspects were discussed in Chapter 4.

Confinement is the key to minimizing the environmental impact to nontarget species. Plant field tests to date have used removal of reproductive structures, the lack of non-cross-pollinating weedy relatives, and distance from related cross-pollinating varieties to prevent new genes or gene combinations from escaping beyond the control of the experiment. Established conditions for confinement of classically modified plants in field tests are being used to limit movement of genes outside the test site, thereby minimizing effects on natural and managed ecosystems. All modified plants that have been field-tested or are proposed for field research are highly domesticated with an established history in field testing.

Potential for Weediness

As discussed in Chapter 4, there are two major issues regarding weediness: (1) does the modified crop itself have weedy properties, and (2) does the modified crop have traits that if transferred to wild relatives would increase their weediness?

The properties generally attributed to weediness include seed dormancy, long soil persistence, germination under diverse environmental conditions, rapid vegetative growth, high seed output, and high seed dispersion (Baker, 1974). These properties are usually thought to represent complicated, multigenic traits. Although it can be argued that only a few genes in certain crops separate them from weeds, crops derived by molecular methods are no more likely to evolve into weeds than crops produced by classical methods.

The introduction of herbicide resistance into crops is receiving research attention. Several crops such as tobacco, tomato, and oilseed rape that have been modified to resist active ingredients of herbicides, such as glyphosate, bromozynil, sulfonylureas, and phosphinothricin, have been tested in the field. The benefit of such research will be increased flexibility in weed control, including benefits such as improved weed control efficacy, reduced costs to farmers, the opportunity to replace currently used chemicals with more environmentally friendly chemicals, and the reduction of overall herbicide usage (Boyce Thompson Institute, 1987).

Herbicide-tolerant plants have been feared to be able to develop into volunteer weeds or to spread resistance genes to weedy species. The key to evaluating that risk is to focus on specific products on a crop-by-trait basis. This involves determining the possibility that herbicide-resistant volunteer plants will become weeds in a subsequent year, the potential for introgression of herbicide resistance genes into weedy relatives, and assessing the potential impact of herbicide-tolerant plants on the cropping and weed-control practices of particular geographic regions. Corn, wheat, and sugarbeets are examples of crops that can become volunteer weeds but are controlled in subsequent crops by cultivation and by different herbicide products. A glyphosate-tolerant volunteer corn plant in a soybean field would be controlled by normally used preemergent or postemergent herbicides. Similarly, a sulfonylurea herbicide-resistant wheat plant could be controlled in either rotational crops or on fallow land with today's normal cultural practices. In addition, past experience from breeding herbicide tolerance into crops—such as metribuzin resistance in soybean, atrazine resistance in canola, and acetanilide resistance in corn—have shown that the phenotypes are stable, and these modifications have not increased the weedy characteristics of a given crop. The primary U.S. crop targets into which herbicide tolerance is being engineered are corn, soybean, and cotton; none of these species outcrosses with weedy relatives in the United States or displays significant potential to develop into weeds themselves.

Specific Examples

We cite as examples, field tests of tomato plants containing (1) the *B. thuringiensis* insect-control protein, which in the laboratory killed caterpillar pests such as tomato hornworm, fruitworm, armyworm, and pinworm (Fischhoff et al., 1987), and (2) coat-protein genes from tobacco mosaic virus (TMV), that confer resistance to infection by TMV (Nelson et al., 1988). Scientific evidence available from published reports, expert opinion, and direct experimentation led to the conclusion that the field introduction of tomato plants tolerant to certain insects and viruses would have negligible environmental impact.

1. The genetically modified tomato plants were well characterized.
   - Greenhouse testing confirmed stability of gene expression and inheritance.
   - The plants were free of *Agrobacterium* spp. used for gene transfer.
   - No unusual phenotypes were associated with genetically modified plants.
2. The introduced genes and gene products were well characterized.
   - The vector DNA contained no uncharacterized sequences.
   - The *B. thuringiensis* protein produced in the plant has no effect on beneficial insects or mammals.
   - The TMV capsid protein has no effect on nontarget species.
3. Biological confinement at the test site was readily available.
   - Bt and TMV proteins decompose in the soil.
   - Tomato normally self-pollinates under field conditions and has no cross-hybridizing weedy relative in North America.

In addition to the confinement afforded by the lack of cross-pollination and the absence of sexually compatible weed species, it was possible to physically contain plants at the test site by fencing to discourage seed dispersal by predators. Also, tillage and chemical means were used to destroy volunteer tomatoes. Small-scale tests conducted in Illinois and Florida over the past 2 years have indicated the absence of environmental impact and provided the following new data.

- The introduced insect tolerance and virus-resistance traits were stable and not transferred to tomato plants as close as five feet away through cross-pollination.

- There were no significant differences in the nontarget insect populations collected in and around modified and control plants.
- Field control of caterpillar pests confirmed laboratory and greenhouse results; there was no extension of control beyond caterpillar species.
- Field control of TMV confirmed laboratory and greenhouse results.
- The plants containing the *B. thuringiensis* endotoxin and coat-protein grew normally; there were no indications of any adverse phenotypes such as plants with increased susceptibility to other viral or fungal diseases.

Although much field research is needed to evaluate the performance of insect-tolerant and virus-resistant tomato varieties under different conditions, the preliminary data confirm the predictable behavior of plants modified by molecular methods and tested in laboratory and greenhouse. They also demonstrate that field research methods developed for crops modified by classical methods are also suitable for field research of crops modified by molecular methods.

SUMMARY POINTS

1. Crops modified by molecular methods in the foreseeable future pose no risks significantly different from those that have been accepted for decades in conventional breeding.
2. The evaluation of plants modified by molecular techniques does not represent a unique concern. Under appropriate conditions of confinement, field-test evaluations can proceed with negligible risk.

6
Conclusions and Recommendations: Plants

Numerous ways exist to modify plants genetically. We have grouped them into three broad categories: classical, cellular, and molecular. Classical methods include sexual hybridization, embryo culture (rescue), undirected mutagenesis, and anther and ovule culture. Cellular methods include cell fusion and tissue culture to produce somaclonal variation. Molecular techniques include several methods (such as recombinant DNA and electroporation) that result in specific insertions of defined DNA sequences.

- Methods used for genetically modifying plants include classical, cellular, and molecular techniques. The molecular techniques are the most powerful and precise for incorporating new traits.

Plants genetically modified by classical techniques have been highly beneficial to society during this century. These benefits are expected to continue and grow in the years ahead through the additional applications of recently developed molecular and cellular techniques. Farmers and consumers will be the major beneficiaries of the improved economic productivity that should keep farmers more competitive in the world markets, while improving food, feed, and new plant products through production practices that are compatible with the environment.

- Society will continue to benefit greatly from genetically modified plants.

WHAT DOES PAST EXPERIENCE TEACH US?

About Introductions

Over the past hundred years, plant breeders and other agricultural scientists have accumulated vast experience and information about the introduction of genetically modified plants into the environment. In fact, almost all of the major crops currently grown in the United States have been introduced from foreign sources and further bred in the United States for improved characteristics. New weed species also have been introduced, although most weeds were introduced in early colonial times, 200 to 300 years ago. Our past experience with these introductions leads to two conclusions: (1) Considerable success has been achieved in introducing and commercializing genetically modified crops and other plants, and (2) problems (economic or environmental), when they occur, are usually minor or manageable.

Domesticated species, such as most field crops, pose little direct threat to the environment. Problems that have resulted from introductions have been indirect, such as increased soil erosion, and associated with managed ecosystems, such as farms. These problems can be effectively controlled by altering such farming practices as crop rotation, cultivation, cultivar selection, or choice of herbicides.

Extensive experience has been gained from routine field introductions of plants modified by classical genetic methods. For example, an individual corn, soybean, wheat, or potato breeder may introduce into the field 50,000 genotypes per year on average or 2,000,000 in a career. Hundreds of million of field introductions of new plant genotypes have been made by American plant breeders in this century. There have been no unmanageable problems from these field introductions through the use of established practices.

- Plants modified by classical genetic methods meet the familiarity criterion on the basis of experience with hundreds of millions of safe introductions over decades. Current oversight practices of such field introductions are appropriate, and no additional oversight is required.

About Genetic Modification

The majority of genetic modifications that are being proposed for domesticated crops by molecular methods are similar to those already achieved by classical means. These include resistance to herbicides, pests, drought, and salt as well as compositional changes in the seed or other plant parts. The genes being used to obtain these traits may differ from those used in the past, but so far these genes have introduced traits with which we have considerable experience. Therefore, our experience with the introduction of plants and genetic modifications by classical means is relevant to the introduction of plants modified by newer methods such as recombinant DNA techniques.

Molecular genetic methods differ from classical and cellular methods in that molecular genetic methods involve manipulation of not more than a few genes and their associated regulatory elements. These genes and their elements are usually well characterized, whereas classical and cellular methods may modify many genes but with limited characterization. Most plants modified by the molecular techniques proposed for field research have been modified only by the addition of one or a few characterized genes within the genome of a domesticated plant. In contrast, classical procedures such as sexual hybridization and some cellular procedures such as cell fusion result in the recombination of entire genomes of the two parental cells. Because the specific gene products added by molecular techniques are better characterized than those added by classical procedures, questions about the changes effected in plants modified by molecular techniques can be asked and answered more precisely.

Experience gained so far from field research on molecularly modified plants, as well as from extensive laboratory and greenhouse research, supports the following conclusions:

- Crops modified by molecular and cellular methods should pose risks no different from those modified by classical genetic methods for similar traits. As the molecular methods are specific in terms of what genes are being added, users of these methods will be more certain about the traits they introduce into plants. Dissimilar traits will require careful evaluation in small-scale field tests where plants exhibiting undesirable phenotypes can be destroyed.

About Weeds

The vast majority of introduced wild species are unsuccessful in their new habitats and therefore fail to become established. However, on occasion introduced wild, non-native species can give rise to problems. A better understanding of the process of species establishment would enhance predictability in determining the ultimate success or failure of an introduced wild, non-native plant. If the newly introduced plant poses problems, standard control measures are usually available.

A major issue is the potential ability of genetically modified plants to hybridize with weedy relatives to yield offspring with characteristics that enhance weediness. On the basis of evidence of gene movement between classically bred plants and weedy relatives, this process occurs infrequently, but varies widely among crops. When gene movement from crops to weeds occurs, the weeds become more croplike and compete for the same resources. This phenomenon is not believed to have caused problems in natural ecosystems, but has within managed ecosystems.

- The potential for enhanced weediness is the major environmental concern surrounding the introduction of genetically modified plants. The incidence of enhanced weediness has been extremely low in the past and has been controllable.

CONTROL AND CONFINEMENT OF GENETICALLY MODIFIED PLANT VARIETIES

An array of options exists for confining a plant to its test site. These options include male sterility, removal of reproductive organs, herbicides, insecticides, disinfectants, tillage, water manipulation (flooding and irrigation), isolation from similar plants, photoperiod manipulation, dates of planting and restriction in the number of locations, and physical barriers such as cages.

Initial field tests of classically bred plants are normally carried out with different plants having many different gene combinations. Great care is taken to track modified properties in the plant and to retain properties of interest by avoiding contamination with similar plants grown in the same or nearby fields. The small scale of plots, such as 20 feet long by 3 rows wide, limits the dissemination of a particular plant. Further, repeated and periodic observations help ensure that undesirable genotypes are not propagated further.

Acreage of selected plants is increased over several years, and each year observations are made of both desirable and undesirable traits. Millions of individual plants are tested annually in the United States, and no environmental damage has been documented from their release. Standard confinement practices have been effective in keeping in bounds both poorly domesticated and highly domesticated plants.

- Proven and routinely applied confinement methods include biological, chemical, physical, geographical, environmental, and temporal control, as well as limitation of the size of the field plot.
- The committee could document no case of escape of a plant introduction from a confined experimental field plot (1 acre or less) that has produced problems in natural ecosystems.

Confinement of plants in small field tests is almost always successful. Particular traits may be transmitted by pollen to other related plants in the vicinity, but most available evidence shows this to be rare. Unless that trait confers a strong selective advantage to the progeny of the recipient, no adverse effects will occur. If necessary, such plants can be destroyed. Molecular genetic techniques neither enhance nor decrease the probability of occurrence of such genetic transfer relative to classical methods of genetic manipulation.

- Established confinement options are as applicable to field introductions of plants modified by molecular and cellular methods as they are for plants modified by classical genetic methods.

LARGE-SCALE INTRODUCTIONS AND COMMERCIALIZATION

This report addresses small-scale experimental introduction, not large-scale introductions and commercialization. As experience with small-scale field research increases, the information gained will permit large-scale introductions to be approached with greater confidence. A plant that exhibits desirable traits in small-scale tests will be gradually increased in number and either returned to a breeding program or released as a cultivar. Oversight mechanisms should remain flexible to accommodate the transition that will occur as testing of crops modified by molecular methods proceeds from isolated field plots to large-scale, multisite testing. This field research is critical to the development of new crop varieties and hybrids.

An important consideration in the commercialization of plants produced by any of the available techniques is how to deploy resistance genes so that they remain effective in controlling pests. For example, if the *Bacillus thuringiensis* endotoxin gene is widely used without regard to the development of resistance, tolerance to this useful bioinsecticide might develop. The same is true for disease-resistance genes.

Another concern associated with large-scale use is the potential genetic "contamination" of populations of wild relatives of cultivated plants by genes from unrelated organisms that have been introduced into the cultivar. However, for this concern to be valid, gene introgression would have to occur and the resulting progeny would have to have a selection advantage in their wild environment.

- Experience gained through small-scale field research is crucial to the large-scale use of genetically modified plants.

A FRAMEWORK FOR ASSESSING RISK

Most of the extensive past experience on field research of plants that have been genetically modified by classical techniques is relevant to field research of plants modified by molecular and cellular techniques. Procedures of confinement, monitoring, and mitigation work equally well, regardless of how the plant was produced.

The types of modifications that have been seen or anticipated with molecular techniques are similar to those that have been produced with classical techniques. No new or inherently different hazards are associated with the molecular techniques. Therefore, any oversight of field tests should be based on the plant's phenotype and genotype and not on how it was produced. The power of the molecular methods, however, does present the possibility that plants with unfamiliar but *desired* phenotypes may be produced. In some cases, new genes sources may be used, but familiar phenotypes will result. Plants with unfamiliar phenotypes should be subject to oversight until their behavior is predictable and shown to be nondetrimental to the environment.

In this section, the committee proposes a decision-making framework (Fig. 6-1) that allows experimental field testing based on (1) familiarity with the plant and genetic modification (Fig. 6-2), (2) the ability to confine the plant (Fig. 6-3), and (3) the perceived environmental impact if the plant should escape confinement (Fig. 6-4).

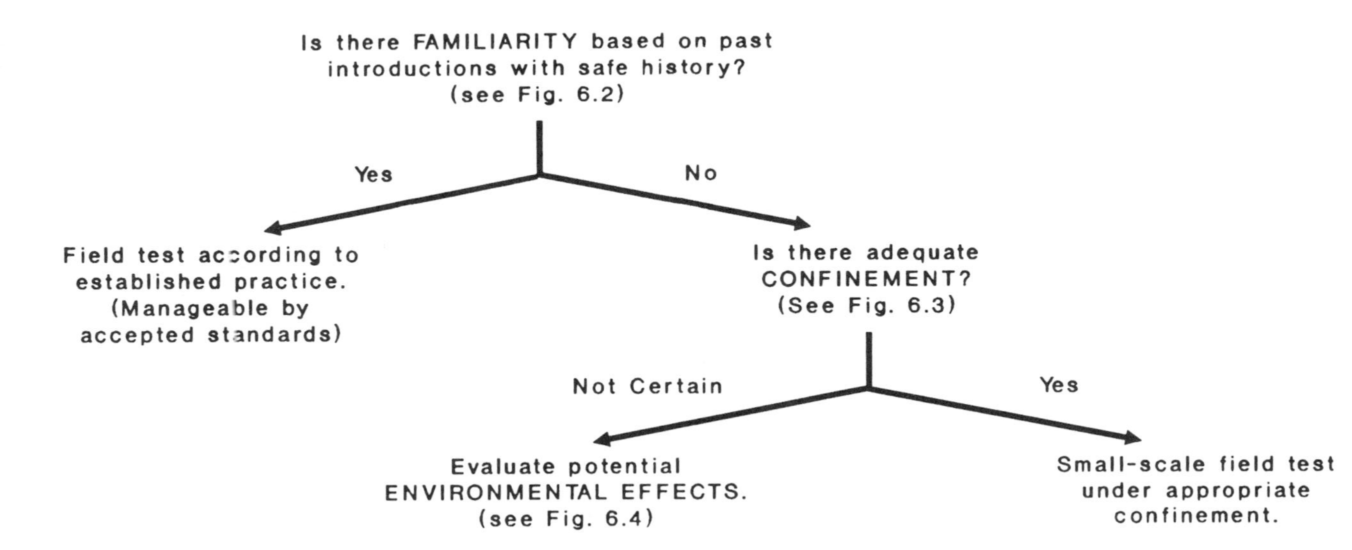

FIGURE 6.1 Framework to assess field testing of genetically modified plants.

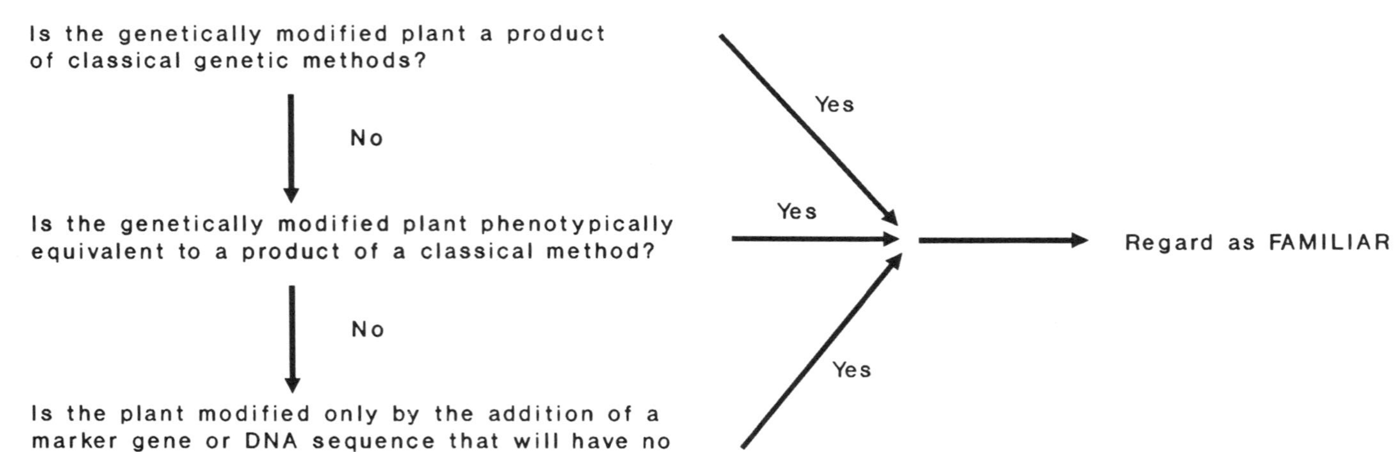

FIGURE 6.2 Familiarity.

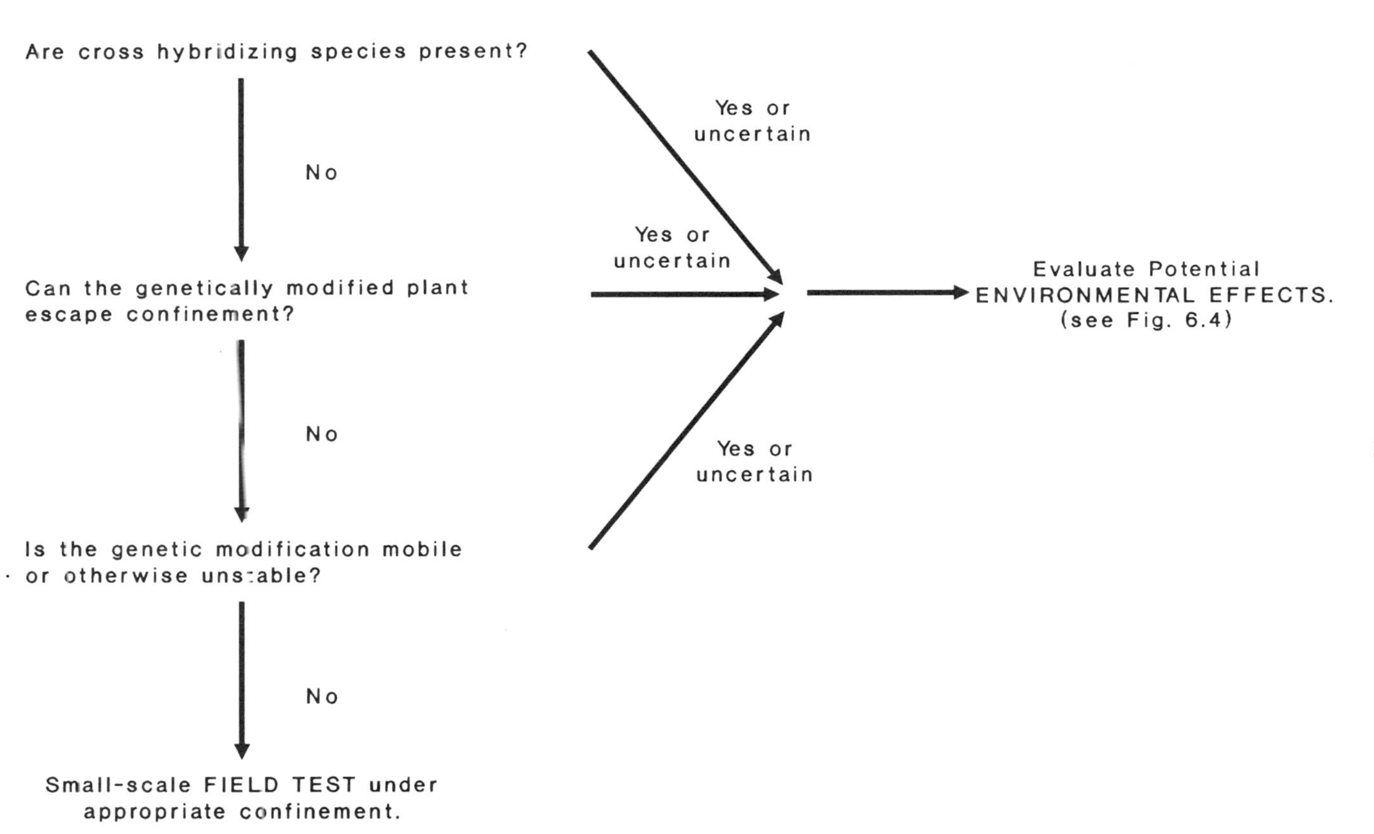

FIGURE 6.3 Confinement.

Situations that are familiar and considered safe on the basis of past experience or experimentation should be classified as manageable by accepted standards (MAS). MAS plants would include, for example, classically produced plants and other plants with familiar phenotypes. These plants should be field-tested in a manner that is most appropriate based on past experience in traditional plant breeding.

All plants can be confined, some more readily than others. The use of sterile plants is probably the best example of easy confinability, providing that attention is paid to the dissemination of vegetative propagules. The other extreme would be to confine an open-pollinated plant in the presence of cross-hybridizing wild relatives. In this situation, confinement may be as strict as physical containment in a quarantine greenhouse. It is clear that the appropriate level of confinement depends on the plant and the geographic area for the field test. If confinement is difficult or uncertain, attention needs to be given to the potential environmental impact of the introduction. If there is potential for considerable negative environmental impact, confinement procedures should be rigorous, as with screened cages. If potential impact is low, less stringent procedures should be called for.

As data based on field tests accumulate, it may be desirable to lessen the confinement requirements so that a plant can be used in a crop-improvement program. Field-test results need to be assessed for potential negative environmental impact as a result of altered characteristics of weediness, toxicity, or pest resistance. Data obtained through field testing provide the best way to assess the presence of undesirable characteristics accurately.

The committee has also included a set of example questions (Figs. 6-1 through 6-4) that might need to be asked at each phase in the decision-making process. This is not a comprehensive list. The importance attached to each of these questions should be determined on a case-by-case basis.

GEOGRAPHIC FRAME OF REFERENCE

Even though the issues discussed in this report are of interest worldwide, it is important to keep in mind that it is written to address the concerns of the United States. The occurrence of weedy or wild relatives for major crops depends on geographic area. In addition, some uses of genetically modified plants will be better

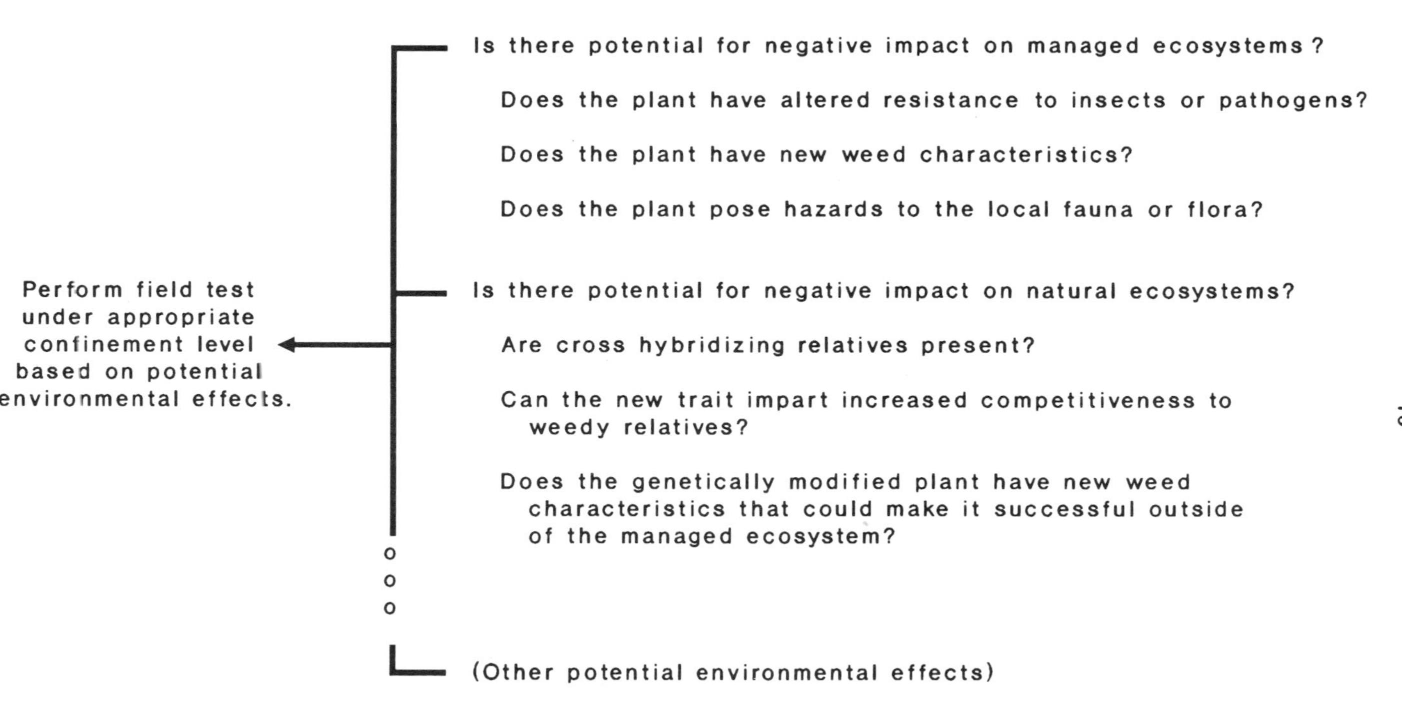

FIGURE 6.4 Potential environmental effects. Appropriate questions for specific applications to be added by users of the framework.

suited for certain geographic areas than others. Therefore, each country and geographic area will need to determine the extent to which the introduction of a genetically modified plant will have an impact on its environment.

OVERSIGHT CONSIDERATIONS

It is reasonable to ask what type of oversight is needed to protect the public welfare and environment, yet does not unnecessarily restrict biotechnological research and commercial development. Institutional or local review has a long history of success. For example, in some cases crop certification of plants for expanded field tests and commercialization uses established institutional review or no review. Similarly decentralized local oversight is also used for approval of animal, human, and other experimental procedures including experiments utilizing recombinant DNA techniques. Therefore, local oversight is seen as a suitable option for assessing the risk associated with most research. It is desirable to delegate low-risk introductions—probably 95 percent or more of all ongoing research—to local oversight, so that federal oversight can be used for those few cases where perceived risk exists because confinement is questionable.

We also note that organizations such as government and university research groups are underrepresented in field introductions of plants modified by molecular techniques. This absence probably arises from the complexity and cost of seeking regulatory approval. This research community has played a vital role not only in developing contemporary procedures and techniques, but also in most plant improvement. If the full benefit of molecular modifications is to be realized, the academic and governmental research communities must have access to and be encouraged to apply the technologies of cellular and molecular genetic modification to plants and to evaluate them through field testing.

7
Past Experience with the Introduction of Microorganisms into the Environment

Microorganisms have had many beneficial applications in agriculture, waste treatment, and food production. In this chapter we describe these and other uses and discuss the prospects for future beneficial applications in clean-up of toxic wastes (bioremediation), mining, and mineral recovery. As discussed in Chapter 2, modern methods of producing novel genetic combinations in microorganisms can be very precise. This history of safe use and an understanding of genetic modification through modern methods combine to give scientists a degree of familiarity with certain microorganisms. This chapter will describe the familiarity we have with certain microorganisms as background for use of a familiarity criterion in a framework to help in evaluating the relative safety or risk of field testing genetically modified microorganisms. We also identify the scientific issues that should be considered when an application involves an unfamiliar microorganism. These issues are then dealt with in greater detail in Chapters 8, 9, and 10.

HISTORY OF BENEFICIAL USES OF MICROORGANISMS AND PROSPECTS FOR THE FUTURE

Food Production

Naturally occurring microorganisms with specialized or unique properties have been used for centuries in food production. The basic microbiology of bread-making has remained substantially the same for thousands of years. Egyptian bakers as early as 2100 B.C. obtained their yeasts from the settlings of beer vats, whereas the Greeks and Romans used yeasts from wine vats, and later the English used brewer's yeast ("barm") (Ayres et al., 1980). Sourdough bread has been produced in the San Francisco area for more than 100 years, with yeasts and a sourdough bacterium that ferments maltose (Ayres et al., 1980). Other fermentations are used in producing pickles, olives, and sauerkraut. Virtually every human culture that utilized domesticated milk-producing animals also developed fermented milk products.

Food production, including baking, brewing, and fermenting foods, is hygienic but not aseptic. Although microorganisms from food-processing industries enter the environment, they have been used safely for centuries. The microbial flora of a food may include microorganisms found on the raw material, those added during processing, and those surviving preservation, treatment, or storage. The textures, flavors, and aromas of foods often depend on a defined mixture of microorganisms, such as those which develop the flavors and aromas of cheeses (Banwart, 1979).

Many opportunities exist to apply genetic modification technologies to microorganisms used in the food industry. For example, the use of classical genetics to improve brewer's yeast is limited because the strains of *Saccharomyces cerevisiae* cannot be crossed; genetic modification by molecular methods provides a valuable alternative strategy for strain improvement (Timmis et al., 1988). Brewer's yeast is not capable of degrading starch, and the amylase needed in brewing is produced by germinating barley. A suggested target for genetic modification of brewer's yeast is construction of amylolytic yeast strains. In addition, enzymes that degrade dextrins generated by the barley amylase could be useful, and the glucoamylase gene from *Saccharomyces diastaticus* has been cloned and expressed in brewer's yeast.

Agriculture

The history of introduction of naturally occurring microorganisms into the environment for agricultural purposes provides extensive data on the release of genetically modified microorganisms. The microorganisms that have attracted the greatest attention for agricultural applications include: (1) *Bacillus thuringiensis* (the source of Bt toxin) and baculoviruses used as insecticides; and (2) bacteria that fix nitrogen in soil and rhizobia that fix nitrogen in association with leguminous plant roots.

Baculoviruses have been used since the nineteenth century to control insect pests. They specifically infect arthropods, do not pollute the environment, and are safe to handle. More than a dozen baculoviruses have been used commercially to control insect pests without any evidence of harm to the environment (Podgwaite, 1985; Entwistle and Evans, 1985, Hunter et al., 1984). Recent studies (Bishop, 1986, 1988; Bishop et al., 1988) demonstrate the suitability of genetically marked baculoviruses for release in field tests. The improvement of baculoviruses by genetic modification has focused primarily on increasing their speed of action and altering their ability to survive over long periods.

Bacteria, such as naturally occurring strains of *B. thuringiensis*, have also been used for years as biological control agents. Toxins in commercially produced microorganisms are effective insecticides and are used on several agricultural and forest pests. The toxins from different strains are specific to certain lepidopteran, dipteran, and coleopteran insect pests (Hofte and Whitely, 1989). The genes conferring production of different *B. thuringiensis* delta endotoxins (Bt toxin) have been cloned and partially characterized (see Lindow et al., 1989, and references therein). Because *B. thuringiensis* does not multiply on plants, effective insect control requires repeated applications. At this time, widespread use of the Bt endotoxin for biocontrol of insect pests has not had any adverse environmental or health effect.

Recently, there have been attempts to overcome the spatial and economic limitations of foliar applications of *B. thuringiensis* by introducing the delta endotoxin gene into the chromosome of *Pseudomonas fluorescens*, an effective colonizer of corn plants (Obukowicz et al., 1987). The organism has not been field-tested but laboratory studies indicate that this genetically modified bacterial strain does not differ from its parental strain in survival and dispersal characteristics, and it is somewhat toxic to root cutworm but not to the corn rootworm.

Transfer of the Bt endotoxin gene into internal colonists of plants, such as *Clavibacter xyli* subspecies *cynodontis* found in the xylem elements of Bermuda grass, shows promise for application in corn (Lindow et al., 1989). Field studies have been initiated with this genetically modified microorganism for control of leaf- or stem- feeding lepidopteran larvae.

Leguminous crops form symbiotic associations with host-specific *Rhizobium* and *Bradyrhizobium* species that fix nitrogen in a form that can be used by the plant, thus increasing agricultural yield. These nitrogen-fixing bacteria have attracted considerable attention with respect to genetic improvement (Schmidt and Robert, 1985). Soil is the major reservoir of free-living rhizobia; the organisms proliferate in the root-nodules of their host plant, and organisms from the decomposed nodules persist in the soil (Nutman, 1975). Due to the persistence of existing rhizobia, any new rhizobia applied to the soil as seed inoculants must function and compete successfully in an extremely dynamic and complex ecosystem (Dowling and Broughton, 1986). Therefore, the most spectacular increases in agricultural productivity from inoculation with a new rhizobium have occurred when rhizobia specific to a certain leguminous plant have been absent from soil before introduction of that legume as a new crop, as in introduction of soybeans into the United States and Eastern Europe and the introduction of pasture legumes into Australia. Commercial legume inoculants have been produced for almost a hundred years, and these rhizobia have been added to soil in enormous numbers without causing undesirable effects. The success of the use of *Rhizobium* spp. inoculants over many years provides a good example of the safe use of genetically modified microorganisms.

Improvements in biological nitrogen fixation will be important for meeting future demands of agriculture, reducing requirements for fertilizer, conserving fossil fuels used in producing and applying nitrogenous fertilizer, as well as in minimizing adverse environmental effects from run-off of nitrogenous fertilizers into lakes and streams. Modern biotechnology will allow greater precision in tailoring microorganisms such as the rhizobia.

Waste Treatment

The most extensive and intensive application of microorganisms released into the environment is in domestic waste treatment, where they are used to reduce the biological oxygen demand and often to reduce the toxicity of sewage effluents. Sludge digesters, settling ponds, trickling filters, and enhanced degradation systems depend on microbial processes. Sewage sludge from large digesters, when pumped into an evaporation pond, represents a massive release of microorganisms into the environment. Yet effluent from a properly operated activated sludge processor or trickling filter poses neither

public nor environmental health problems. The history of releasing treated sewage effluent into the environment argues convincingly that these procedures are safe.

Bioremediation

Biodegradation of pollutants in the environment has used both naturally occurring as well as genetically modified microorganisms (Bopp, 1986; Brown et al., 1988; Focht, 1988; Frantz and Chakrabarty, 1986; Frick et al., 1988; Ghosal et al., 1985; Kilbane et al., 1983; Senior et al., 1976; Unterman et al., 1987). Nonbiological methods, such as physical confinement, chemical degradation, or incineration, are costly and may disperse toxic compounds or their derivatives into the environment. Biodegradation therefore offers a valuable alternative (Kobayashi and Rittmann, 1982). In situ degradation by introduced microorganisms represents a practical solution to certain types of environmental pollution. The attractiveness of bioremediation treatment is that in both surface and subsurface environments microbial processes may permanently remove organic contaminants, rather than merely contain them. In addition, bioremediation by introduced microorganisms may continue in situ even after introductions have ceased. Genetic tools can be used to enhance specific catabolic pathways of biodegradation of foreign compounds by microorganisms under a wide range of environmental conditions (Lindow et al., 1989).

Trichloroethylene (TCE) is the contaminant on the Environmental Protection Agency list (1985) most frequently reported at hazardous waste sites. TCE and other chlorinated alkenes present a serious groundwater contamination problem because they are suspected carcinogens that resist biodegradation in the environment (Infante and Tsongas, 1982). Unfortunately, in anaerobic subsurface sediments and aquifers, chlorinated alkenes can be converted to even more powerful carcinogens, such as vinyl chloride (Bouwer et al., 1981; Wilson et al., 1983). However, an aerobic microorganism will degrade TCE when the organism is simultaneously exposed to phenol (Nelson et al., 1986). Recently, an aerobic, methane-oxidizing bacterium that degrades TCE has been isolated; in pure culture it degrades TCE at concentrations commonly observed in groundwater (Little et al., 1988). Destruction of TCE by microorganisms in contaminated groundwater has also been demonstrated (Fliermans et al., 1988). Monooxygenase genes have been cloned into a strain of *Escherichia coli*, and this genetically modified organism has been shown to degrade TCE completely (Winter et al., 1989).

Polychlorinated biphenyls (PCBs) are ubiquitous pollutants that are physically inert and poorly soluble in water. Although PCBs were believed to be refractory to biodegradation, many PCBs containing fewer than four chlorines per molecule can be degraded biologically (Focht and Brunner, 1985; Rochkind-Dubinsky et al., 1987; Shields et al., 1985). Field studies on degradation of PCB in soil treated with nutrients and inoculated with PCB-degrading bacteria have been conducted near Oak Ridge, Tennessee. Biodegradation increased markedly in the inoculated soils. Gene-probe analyses and growth of the microorganism in culture demonstrated that the added microbial strain survived (Hill et al., 1989). Anaerobic degradation of PCBs has been demonstrated by Quensen et al. (1988). Genes involved in PCB degradation have been cloned by Taira et al. (1988) and Kimbara et al. (1989), an accomplishment that should allow future development of new genetically modified PCB-degrading microorganisms.

Investigators have sought, by genetic modification, to combine several degradative steps in a single microorganism to be used for destroying pollutants (Ghosal et al., 1985; Tomasek et al., 1989; Focht, 1988; Sangodkar et al., 1988; Golovleva et al., 1988; Timmis et al., 1988). Timmis et al. (1988) tested biodegradation of substituted aromatic compounds (chloro- and methylaromatics) in activated sludge; modified microorganisms survived and functioned with no adverse effects on the rest of the microbial community and with no transfer of the new genetic information to indigenous microorganisms. Genetic modifications also may provide significant improvements in the rates of degradation and the range of toxic pollutants subject to degradation.

Mining and Mineral Recovery

The principles of microbial biogeochemistry (Ehrlich, 1981) and the potential for recovering minerals through improved microbial processes may be exploited within the next decade. Thiobacilli are autotrophic and can derive their energy from the oxidation of sulfur compounds or ferrous iron. Techniques for manipulating the genes of this group of microorganisms to enhance leaching of copper and uranium are being developed (Yates et al., 1988). Although genetic technology as applied to this group of microbes is less advanced than that for other microorganisms, plasmids have been constructed that may enhance the recovery of gold from ores by *Thiobacillus ferrooxidans* and increase the organism's resistance to arsenic compounds (see Lindow et al., 1989, and references therein).

Commercial bioleaching operations in the mining industry represent another application of microorganisms—treatment of ores in heaps or pits. Adverse environmental effects have not been reported from the introduction of improved bacterial strains for mining applications of biotechnology (Nicolaides, 1987).

SIMILARITIES AND DIFFERENCES BETWEEN CLASSICAL AND MOLECULAR METHODS

The preceding section illustrates a long history of beneficial applications of microorganisms in food processing, agriculture, waste treatment, and bioremediation. In many instances, such applications can be performed even more effectively with microorganisms that have been genetically modified, by either classical methods or molecular techniques (NAS, 1977; OTA, 1984, 1988; Gillett et al., 1985; Korwek and de la Cruz, 1985; Olson, 1986; Timmis et al., 1988). Among the products of classical methods of genetic modification are spontaneous mutants and many recombinants formed by natural processes such as conjugation, transduction, and transformation in bacteria. As mentioned in Chapter 2, site-directed mutagenesis, DNA cloning, and cell fusion are included among the molecular techniques.

Why then employ the recently developed molecular techniques to produce microorganisms to be introduced into the environment? First, the newer methods permit greater precision in the construction and characterization of the desired genotypes. Enhanced precision means scientists know more exactly the changes that have been effected and can make better judgments about the safety of genetically modified organisms (Brill, 1985; Davis, 1987; Tiedje et al., 1989). Second, molecular methods (either alone or in conjunction with classical approaches) may permit the formation of novel combinations from distantly related genomes, combinations that would be extremely difficult or even impossible to obtain by classical methods. In essence, molecular techniques allow scientists to bypass natural barriers to genetic exchange (Tiedje et al., 1989; Wong et al., 1988, Jaynes et al., 1987).

With the precision and power of new techniques, researchers have begun to develop new genetically modified microorganisms in an attempt to enhance and extend beneficial applications (Orser et al., 1984; McCormick, 1985; Obukowicz et al., 1987; OTA, 1988; Gillett et al., 1985, and references therein; Olson, 1986; Korwek and de la Cruz, 1985; NAS, 1977; Timmis et al., 1988; Lindow et al., 1989). These genetically modified microorganisms can be used in the production of food, pharmaceuticals, and industrial chemicals as well as in agriculture, pollution control, and mining.

CONSIDERATIONS ASSOCIATED WITH UNFAMILIAR APPLICATIONS OF MICROORGANISMS

It is apparent from the examples in the preceding sections that microorganisms have already provided many benefits when introduced into the environment and that molecular techniques to modify microorganisms hold great promise for enhancing and extending these benefits. The experience gained from the use of microorganisms over the years provides scientists with considerable familiarity as they design certain types of field tests. For example, we know from experience that many microorganisms can be used safely in the environment even if we are uncertain of their precise roles in the microbial community or ecosystem.

Were our knowledge of the genetically modified microorganism and the environment perfect, we could assess all risks precisely and make a thoroughly informed judgment based on the magnitudes of the anticipated benefits and potential costs. Because we lack perfect knowledge, we must base our predictions of responses to the release of genetically modified microorganisms on sound scientific inference and on a long record of safety. Microbial ecology is a rapidly developing field that has the potential to add greatly to our knowledge of how to ensure the safe and effective environmental application of microorganisms.

Unfamiliar microorganisms or instances of substantial uncertainty about the interaction of the microorganism and the environment into which it is to be introduced require careful evaluation prior to field testing. Issues pertaining to the biology of a field test that have ecological significance include the following: the functional role or niche of the microorganism in the microbial community and ecosystem, the potential for gene exchange between microbial taxa, the ability to monitor persistence and spread of microorganisms, the potential ecological consequences of the persistence and spread of such microorganisms, and the potential measures to control, if necessary, the effects of introduced microorganisms. In Chapters 8 through 10, we examine these and other issues that may be sources of ecological uncertainty for some microbial field tests.

SUMMARY POINTS

1. Microorganisms have a long history of use in food production, agriculture, and waste treatment. Many opportunities exist to use genetically modified microorganisms to enhance these and other applications, including the clean-up of environmental pollutants and the recovery of minerals. Familiarity with particular microorganisms, their functions, and their target environments is important to consider in assessing potential environmental effects. Familiarity has been incorporated as the first criterion in the framework presented in Chapter 11 for the evaluation of the safety or risk of field-testing microorganisms.

2. Both similarities and differences exist between classical and molecular methods for genetically modifying microorganisms. Comparable modifications can often be accomplished with either method, and each type follows procedures to selectively enrich those genotypes that have the desired phenotypic properties. Molecular methods often provide greater precision in generating the desired genotype and greater power in producing novel genetic combinations.

3. Much has been learned from past experience with microorganisms, such as those used as biocontrol agents or for nitrogen fixation, and the information provides a basis for assessment of relative safety or risk. It may be difficult to assess all potential risks precisely, especially for unfamiliar microorganisms. Although some uncertainties may persist, they can be resolved scientifically as our knowledge of microbial ecology increases.

8
Properties of the Genetic Modification

Powerful new molecular methods for DNA manipulation provide a means for constructing microorganisms with novel genotypes that cannot be duplicated by classical methods and would be highly unlikely to occur naturally. In addition, these new methods enable genotypes to be characterized with a degree of precision not previously available. Genes now can be manipulated by methods for the synthesis, sequencing, cutting, and splicing of DNA molecules, and changes often can be characterized to the base sequence of the DNA. Unmodified or modified, genes can be inserted into virtually any microorganism, even if the donor and recipient microorganisms do not exchange genetic information under natural conditions. Molecular methods of DNA modification also permit deletion of a precise region of a genome and its replacement with a similar or altered segment of DNA.

Thus, the new molecular methods support the production of novel genetic combinations so that microorganisms can perform new functions. The question has been raised whether the microorganisms produced by these methods present any risks not associated with microorganisms produced by classical microbial genetic techniques.

The fact that many molecular methods provide far finer precision than has previously been available means that scientists can produce modified organisms with predictable genotypes. A general example will illustrate the superiority of the precision of molecular genetic modification over classical genetic modification. In classical procedures, to transfer the ability to produce a particular protein from one bacterium (the donor) to another (the recipient), *total* DNA from the donor would be used to transform the recipient. Selection for the desired trait would yield a strain that produced the target protein, but also could contain additional DNA that contributed to the recipient's phenotype. Through the use of new molecular procedures, the target gene can be cloned from the donor and used *alone* to transform the recipient to produce a strain differing from its parent in only one gene. Precise knowledge of genotypic modifications would simplify experiments to determine the nature of the phenotypic changes.

HISTORY

To date, only a relatively few field trials have been conducted with microorganisms modified by molecular methods. It is instructive to review three of these.

Ice-Nucleation-Deficient *Pseudomonas* Mutants

Certain epiphytic pseudomonads damage plants by acting as nuclei for ice-crystal formation. To control this problem, it has been suggested that pseudomonads unable to form ice nuclei might protect plants by excluding colonization by ice-positive (ice^+) strains; both ice^+ and ice negative (ice^-) pseudomonads occur naturally on plants. The ice^- mutation in *Pseudomonas syringae* was produced by deleting from an ice^+ strain the specific gene encoding a protein that acts as the nucleus for crystallization (Lindow, 1988). The entire nucleotide sequence of this chromosomal gene is known, and, the protein has been isolated and well characterized. The ice^- deletion mutant has been characterized genetically and physiologically (Lindow, 1988). The modified genome contained no detectable foreign DNA sequences; it no longer produced the ice-nucleation protein; it did not induce ice nucleation in vitro; and it was phenotypically indistinguishable from its parent in all other traits examined. It should be noted that it was selected for increased colonization, and changes in this characteristic were found. In field tests, the organism performed as anticipated (Lindow and Panopoulos, 1988); that is, it reduced ice-nucleation on plants to which it was applied, it showed no detectable spread beyond the test zone, and after 18 days it could not be detected in the soil.

Transfer-Defective Mutant of *Agrobacterium radiobacter*

Agrobacterium radiobacter is closely related to *Agrobacterium tumefaciens* and is a common soil inhabitant. Strain K84 is used commercially to control crown gall, a bacterial disease of plants. It does this in part by producing a highly specific antibiotic, called agrocin 84, that kills susceptible pathogens (New and Kerr, 1972; Kerr and Htay, 1974; Ellis et al., 1979). Production of the antibiotic, as well as immunity to the agent, are encoded by a plasmid present in strain K84 (Ellis et al., 1982; Slota and Farrand, 1982; Farrand, et al., 1985). This plasmid is self-transmissible (Tra^+), and its transfer to pathogens could result in their becoming immune to control by strain K84 (Panogopoulos et al., 1979). Molecular methods were used to delete from the plasmid the genes required for conjugal transfer (Jones et al., 1988). The genetic alteration occurred as anticipated, and no foreign DNA sequences were detected in the new strain. The deletion had no effect on production of the antibiotic by strain K84, and, aside from the Tra^- character, the altered strain was phenotypically indistinguishable from its parent. In field tests conducted in Australia, the altered strain controlled crown gall as well as its unaltered parent (Jones and Kerr, 1989). Moreover, because it can no longer transfer the plasmid, it is considered a safer strain for long-term use in the field.

Lactose-Catabolizing *Pseudomonas fluorescens*

The field use of fluorescent pseudomonads to control certain plant diseases has generated much interest (Davison, 1988). As a tool for the study of how such organisms behave in the environment, one such pseudomonad was marked with a genetic trait that would allow its direct selection from environmental samples. The marker chosen was the ability to utilize lactose; the gene for lactose utilization is rarely found in species of *Pseudomonas*. The genes for lactose utilization from *Escherichia coli* were inserted into the chromosome of the *P. fluorescens* strain with a transposon delivery system (Barry, 1986; Drahos et al., 1986). Transposons are mobile genetic elements that cause mutations by interrupting the DNA sequence of genes into which they insert. They sometimes act by preventing expression of genes distal to the insertion site, a phenomenon called polarity. Physical analysis confirmed the insertion, but the exact site was not known. The marker no longer was transposable, however, because the transposon was defective. The altered strain was indistinguishable from its parent except for its associated ability to catabolize lactose and thus to form blue colonies on an agar medium containing an indicator dye. The strain has been used in field tests to determine the efficacy of the *lac* marker for tracking purposes. Reports to date indicate that the marker works well for its intended purpose and that the *P. fluorescens* strain shows no unexpected behavioral patterns in the environment (Drahos et al., 1988).

From these three examples, two key conclusions can be drawn. First, microorganisms can be engineered with the new molecular methods such that the genetic alterations are known precisely. Second, these modified microorganisms have behaved in the environment as anticipated. That is, the genetic alterations have produced no documented phenotypic alterations, either desirable or detrimental, that had not been predicted.

GENETIC CONSIDERATIONS

A number of important considerations arise with respect to the actual genetic alteration of microorganisms destined for release into the environment. These considerations apply to manipulations carried out by classical microbial genetic techniques as well as to those performed with new molecular procedures.

Type of Genetic Alteration

Two major types of alterations should be considered. The first is removal of a trait, either by mutational inactivation of the encoding gene or genes or by deletion of the DNA region encoding these determinants. The second is the addition of new traits to an organism. Additions can be accomplished by inserting new genes into the chromosome or indigenous plasmids of the organism or by introducing a new plasmid encoding the traits of interest.

These types of manipulations connote nothing with respect to the sources of the genes or the particular methods used. Manipulated genes can be from a substrain of the organism, from a near or distant relative, or from some unrelated organism. For any of these sources, the alteration potentially could be effected by classical microbial genetic techniques, new molecular methodologies, or some combination of the two.

For mutational inactivation, commonly used techniques involve insertions of transposons. These elements often encode resistance to one or more antibiotics, and expression of these resistance functions acts as a marker for the presence of the elements. Use of transposons has two drawbacks with respect to environmental releases. First, they are mobile and can be excised from their original target site, thus reversing the mutation. Second, their mobility allows them to move to other sites or, under proper conditions, to other organisms. The potential spread of such antibiotic resistance determinants has led many scientists to suggest that alternative methodologies for insertional mutations be used. Two solutions are available: Certain transposons have been modified to render them incapable of excision or further transposition after insertion at the site of interest (Barry, 1986). Alternatively, "gene cassettes" can carry fragments of DNA for insertion in sites within genes of interest (Close and Rodriguez, 1982; Prentki and Kirsch, 1984). Such insertions disrupt gene function in a manner similar to that of transposons, but they have the advantage of producing stable genetic alterations.

It has also become possible to introduce precise mutations by changing bases at predetermined positions within a gene (Taylor et al., 1985). Reintroduction of this modified gene into the host of origin, and its exchange for the wild-type allele, constitutes an extremely accurate method for constructing mutants lacking a particular function. Such alterations can be permanent and do not introduce antibiotic resistance traits. However, the methodology requires detailed knowledge of the target gene.

Mutations that result from insertions or base substitutions are theoretically revertible, even in the absence of genetic exchange. Deletion mutations, on the other hand, are considered nonrevertible because all or part of the gene in question has been physically removed. Perhaps the most useful technology to effect such deletions involves gene replacements (Ruvken and Ausubel, 1981). The gene to be altered is cloned, and a small piece is removed; both steps use molecular methods. Because the cloned DNA is small, the exact nature of the deletion can be designated to the level of the nucleotide sequence. The cloned DNA is then reinserted into the target host and the altered gene is made to exchange for the wild-type copy. This produces an organism containing the altered gene, lacking the wild-type copy, and expressing the phenotype of a deletion mutant (Scolnik and Haselkorn, 1984; Jones et al., 1988, Lindow, 1988). The technique can be performed in such a way that no other traits, such as resistance to an antibiotic, are added during the process (Jones et al., 1988).

Addition of new genes can be accomplished by introducing them onto new plasmids or by inserting them into the chromosome or indigenous plasmids of the target organism. Introduction of the genes onto a new plasmid will invariably result in the acquisition of any additional traits encoded by that plasmid. The desirability of such ancillary traits must be considered within the context of the intended use of the organism. Plasmids introduced into a target microorganism may be unmodified elements isolated from related or unrelated bacteria, constructs made by classical genetic means, or the products of new molecular methods.

Several strategies can be followed to introduce new genes into chromosomes or indigenous plasmids of target organisms. In one, the genes are cloned into transposons, which are allowed to insert randomly into the plasmid or chromosome of the target organism. This strategy has the same disadvantages as using transposons as insertional mutagens—namely, the mobility of transposons and their ability to be excised. Another strategy is to use defective transposons (Drahos et al., 1986). Third is recombinational marker exchange; conceptually this is similar to gene replacement for producing deletions, as described above. The new genes are cloned into a segment of DNA previously removed from the target organism, and this modified segment then is reintroduced into the target organism. Obviously, the piece of DNA into which the new genes are inserted must be chosen carefully, so that manipulations do not inactivate essential genes. Such deleterious effects usually can be minimized by cloning new genes into dispensable or repeated sequences.

Regulation of Gene Expression

Two broad classes of genetically modified microorganisms have been discussed: ones in which function is lost because a gene has been deleted or inactivated (for example, construction of ice$^-$ pseudomonads) and ones acquiring a new function because a gene has been added (for example, production of an insecticidal toxin). In the former case, there is no concern about expression of the gene if the deletion or the inactivating event is permanent. With an added function, molecular methods may allow precise control of the expression of that function.

Expression of most bacterial genes is regulated at the level of RNA transcription (see, for example, Reznikoff and Gold, 1986). Genes are turned on by promoting or permitting their transcription, and they are turned off by preventing their transcription. Based on extensive work with *E. coli* (and to a lesser extent with *Pseudomonas*, *Rhizobium*, and *Bacillus*), it is known that a number of systems are available for controlling the expression of cloned genes. Some systems can be turned on, in the laboratory, by the addition of inducer molecules. For example, a gene in *E. coli* controlled by the *lac* promoter can be switched on with isopropyl β-D-thiogalactopyranoside, a lactose analog. Others can be turned on by ultraviolet irradiation or by heating. In *Rhizobium* species, it is possible to induce gene expression by adding secondary plant metabolites, such as flavonoids present in root exudates (Firmin et al., 1986; Peters et al., 1986; Redmond et al., 1986; Djordjevic et al., 1987). Similarly, certain genes in *Agrobacterium* strains are specifically induced by phenolic compounds, such as acetosyringone, present in plant wound exudates (Stachel et al., 1985; Stachel et al., 1986).

With inserted genes, the mode of regulation may be especially important. Is the inserted gene under the control of its own promoter? Does the modified microorganism produce the proper regulatory signal, or is the gene for the regulatory component also present in the inserted genetic material?

In some cases, it may be useful to place the gene or genes of interest under the control of a promoter that responds to inducing stimuli in a manner most appropriate for the intended function in the environment. If the microorganism is modified to degrade a chemical pollutant, for example, expression of the degradative functions could be controlled by the presence of the appropriate chemical. Microorganisms designed to interact with plants (for example, to protect them against pests) might be modified so that specific gene functions could be induced by a plant product such as a component of a root exudate.

Under other circumstances, constitutive expression of the modified traits may be advantageous, for example, in organisms designed for generalized functions under a broad range of environmental conditions. Constitutive expression may also be advantageous in organisms designed for specific tasks or habitats. For example, constitutive expression of a biodegradative function might allow maximal performance of biodegradation, while conferring a metabolic burden to put the organism at a competitive disadvantage upon depletion of the target chemicals. Thus, the microorganism might disappear from the ecosystem when its degradative task was complete.

Intended and Unintended Changes

Whether the genetic alteration involves the addition or deletion of functions, it always has been important to determine if traits other than those intended have been affected. One of the great advantages of the molecular genetic techniques is that they are usually more precise, and hence their use minimizes the likelihood of unintended phenotypic changes. Nevertheless, even precise genetic manipulation may produce unintended genetic changes owing to the pleiotropic nature of gene action. For example, changing a single gene may make a bacterium resistant to a bacteriophage while also changing its ability to compete for limited resources. In this example, the surface receptors that determine phage sensitivity often are also involved in transport of nutrients into the bacterial cell (Braun and Hantke, 1977). Similarly, in isolated instances a change in host range may be caused by introducing genes at other than the target locus (Staskawicz et al., 1984; Gabriel, et al., 1986).

Precision of Characterization

Many of the new molecular methods permit a degree of genetic characterization unobtainable by classical genetic exchange and recombination processes. In some instances, it may be important to have a thorough knowledge of the genes being manipulated, even to their base sequences. This knowledge can provide information on regulation of gene expression at the levels of transcription and translation, as well as on the properties, activities, and fates of the gene products. New molecular methods exist to make precise alterations in the genes, to verify these alterations, and to test their effects on the expression of the genes.

This should not be taken to mean that only constructs characterized to base sequence are suitable for environmental release. This degree of characterization certainly is not available for organisms constructed by classical genetic techniques. Furthermore, one can conceive of microorganisms modified by new molecular methods, but having complex genetic systems, in which genetic characterization of the sequence would be exceedingly difficult. Again, we emphasize that it is the phenotypic characterization of the modified microorganism that is of primary concern.

Source of New Genes

Considerable attention has been given to the sources from which new genes are derived for insertion into modified organisms. For example, might the acquisition of one or a few genes from a pathogenic source convert the recipient into a pathogen? This question is addressed in depth in Chapter 9. The overall conclusion is that an unrelated microorganism does not become a pathogen merely because it receives a small portion of the DNA of a pathogenic species. Instead, pathogenesis is a complex phenomenon requiring the coordinated action of many genes (Miller et al., 1989).

Nevertheless, a nonvirulent microorganism could become virulent if it received the necessary genes from a *related* pathogen that complemented the recipient's existing (but incomplete) virulence factors. For example, when a single gene encoding an enzyme that inactivates the pea phytoalexin, pisatin, was transferred from the fungal pea pathogen, *Nectria hematococca*, to the fungal corn pathogen *Cochliobolus heterostrophus* (Olen Yoder, personal communication, 1989), the transformed fungus produced lesions on pea leaves, whereas before it had produced lesions only on the stems of peas. Note that the recipient organism was a pathogen in its own right. Although the new genes increased the pathogenicity of the organism, in that it produced lesions on pea leaves, these genes did not convert a nonpathogen to a pathogen.

Genetic Markers

Marker genes, introduced for tracking or identifying the modified microorganism, can be introduced to evaluate the spread of the organism, its persistence in the intended environment, its ability to colonize a particular habitat, or its capacity to exchange genetic information with indigenous microflora.

Antibiotic resistance traits often have been used for such purposes, but they have drawbacks; resistance functions often impose substantial metabolic burdens (Zund and Lebek, 1980; Lee and Edlin, 1985), and their use as markers may contribute to the proliferation of antibiotic-resistant microorganisms. As alternatives, marker genes encoding enzymes for the catabolism of substrates not normally utilized by the particular microorganism have been used successfully. For example, the *lacZY* genes of *E. coli* were introduced into a soil pseudomonad that could not catabolize lactose. This allowed it to utilize the disaccharide and permitted its differentiation from *lac*$^-$ pseudomonads because of its acquired ability to form blue colonies on plates containing the chromogenic substrate, X-gal (Hemming and Drahos, 1984; Drahos et al., 1986; Drahos et al., 1988: also see above). Other candidate marker genes encoding catabolic functions include *xylE* and opine utilization determinants (Tempé and Petit, 1983; Dessaux et al., 1987).

Synthetic polynucleotides or DNA fragments encoding no functions but inserted into the chromosome or into plasmids can serve as markers. Nucleic acid hybridization with probes homologous to the introduced sequences then serve to identify the marked organisms (Attwood et al., 1988). The polymerase chain reaction (PCR) provides a variant on this strategy (Steffan and Atlas, 1988), in which a specific sequence unique to the organism to be tracked is greatly amplified by cycling through a DNA synthesis system. This amplified DNA is detected by hybridization with probes specifically homologous to the sequence. The probed sequence may be a naturally occurring part of the microorganism's genome, or it may be a segment (synthetic or recombinant) inserted into the genome. However, it must be unique to the organism being tracked. The only other requirement is that techniques must be available for the effective extraction of DNA from the samples to be assayed. PCR methods have the potential for increasing the sensitivity of detection while maintaining a high degree of specificity. However, markers based on hybridization are limited in that they cannot be used for direct selection of the microorganism from environmental samples. Rather, they are useful only for screening samples that may contain a large excess of other organisms.

Biological Confinement

No discussion on genetic alterations would be complete without consideration of genetic manipulations intended to biologically confine the modified microorganism and its genes. Bacteria can and do exchange genetic information in the environment with low frequency (Levy and Novick, 1986; Stotzky and Babich, 1986; Trevors et al., 1987; Schofield et al., 1987). The scientific literature gives limited information on gene transfer in nature under realistic conditions of population densities and structures. However, it is clear that microorganisms can exchange genetic information in soils and water and on, or within, plants and animals. The critical variables, namely the frequencies of such exchanges under natural conditions, are poorly known and warrant further studies with marker genes. Because exchange occurs, it is desirable in some instances to use gene combinations that minimize the possibility that these genes will be transferred to other organisms in the environment. Whenever possible, new genes should be introduced onto the chromosome of the target organism. As an alternative, nonconjugal and nonmobilizable plasmids (Levin and Rice, 1980), either indigenous or introduced, may be appropriate if they are not disseminated to indigenous microorganisms under field conditions. Markers can be used to identify indigenous organisms that have acquired genes from an introduced strain.

Situations exist in which the spread of a modified trait is desirable. For example, consider hypovirulence for controlling a fungal disease such as chestnut blight caused by *Cryptonectria parasitica*. Certain strains of *C. parasitica* have been identified that are considerably less virulent than the pathogens that have destroyed the American chestnut forests. This hypovirulence is associated with the presence of double-stranded RNA (Fulbright, 1989). Inoculation of infected chestnut trees with the hypovirulent strains can effectively halt progression of the blight and allow the tree to overcome the infection. This phenomenon appears to be associated with the transfer of the double-stranded RNA molecules to the virulent pathogens. This induces their conversion to hypovirulence (Fulbright, 1989). The spread of the genetic elements—in this case naturally occurring double-stranded RNA molecules—is a necessary feature of the system. One can envision that genetic modification of the double-stranded RNAs might lead to better control agents.

If a microorganism might persist beyond the intended period of usefulness it may be necessary to utilize confinement, for example, biological confinement by suicide genes (see also Chapter 9). For example, a modified microorganism could carry a "suicide" function that is repressed only when the introduced microorganism is performing its intended function. When that function is completed, the organism is killed by derepression of a suicide gene (Molin et al., 1987; OTA, 1988; Curtiss, 1988). Alternatively, microorganisms intended for environmental introductions might be confined by incorporating into their genomes restrictive nutritional requirements, which could prevent the microorganisms' persistence or spread beyond the intended time or space.

In practice, however, the goal may be to maintain the organism in the environment. This may be hard to achieve at times because competition with indigenous microorganisms may make it difficult for a modified microorganism to persist at a population density sufficient for it to carry out its intended function. Modified strains may lose their environmental competence as a consequence of burdens of carriage and expression of additional functions, or as a consequence of the strain's prior adaptation to laboratory conditions.

SUMMARY POINTS

1. Modifying microorganisms by classical genetic methods may often be possible. In other cases, the development of novel genotypes may require molecular methods. The key concern, however, is not the method by which the microorganism is modified, but rather the phenotypic properties conferred by the microorganism's new and preexisting genomic complements.
2. Molecular methods, used either to modify or to characterize genotypes, provide a degree of precision unavailable through classical microbial genetic techniques. This precision increases our knowledge about the genetic alteration and may improve our ability to predict how the organism will perform in the environment.
3. When evaluating microorganisms for environmental introductions, the following characteristics should be considered: (1) the influence of the genetic alteration on the relevant phenotypes of the organism and (2) the genetic mobility of the altered traits. The presence of foreign DNA in a modified microorganism is, by itself, of little concern, but how it influences the expression of phenotypic traits and the mobility of genetic material is important.
4. The pathogenicity of a microorganism results from a complex interaction among a number of genes and gene products of the pathogen and host. It is highly unlikely that moving one or a few genes from a pathogen to an unrelated nonpathogen will confer on the recipient the ability to cause disease. If the recipient is closely related to the pathogenic donor, increased virulence may result if genes directly affecting virulence are transferred. The phenotypic properties of a microorganism intended for introduction into the environment are of primary concern and the source of the gene has relevance only for understanding the nature of the modification.

5. In initial field testing of unfamiliar genetically modified microorganisms, it may be desirable to mark the microorganism so that it can be monitored after its introduction into the field. It also may be desirable to effect genetic modifications designed to limit persistence and minimize transfer of genetic material to the indigenous microflora.

9
Phenotypic Properties of Source Microorganisms and Their Genetically Modified Derivatives

In this chapter, we shift our attention from characterization of the genetic modifications to characterization of the ecologically important phenotypic properties of the unfamiliar microorganism intended for field testing and eventual introduction into the environment. The most relevant phenotypic properties are those that relate to the persistence of the introduced microorganism (or the genetic material incorporated during its modification) and those that affect the ecosystem. Of course, for a microbial application that is familiar, such phenotypic information is likely to be already available or rendered unnecessary by a safe history of past use in the environment.

PERSISTENCE

Persistence can be viewed as survival of the introduced modified microorganism or retention of particular genetic traits in new genetic combinations resulting from gene transfer.

Persistence of the Microorganism

If an organism cannot persist in a particular environment, it poses little threat of causing prolonged environmental impact. However, short-term responses may be seen, as in the use of *Bacillus thuringiensis* to control lepidopterous insect pests. An organism that is killed or does not persist might be viewed as similar to a chemical treatment that produces no chemical residues. For example, some organisms will be developed to control pests or degrade pollutants and will be used once or reapplied, if necessary. On the other hand, the utility of some applications of genetically modified microorganisms will depend on their ability to persist in the environment. For example, genetically modified microorganisms introduced into the rhizosphere for plant growth promotion or disease control will be most beneficial to farmers if the organisms remain active for years.

Persistence and spread are particularly relevant, however, if a proposed application is both unfamiliar and has some potential for adverse environmental effects. In such cases, it may be difficult to mitigate an adverse environmental effect simply by halting further application. Therefore, the potential for persistence and adverse effects should be considered together when establishing levels of concern for proposed field tests involving unfamiliar microorganisms; this is reflected in the framework presented in Chapter 11. For unfamiliar applications, it is prudent to evaluate potential adverse effects prior to field testing (Vogel and McCarty, 1985; Tiedje et al., 1989).

The persistence of an unmodified microorganism in its usual habitat is largely predictable. Genetic modification could influence persistence if it changed the fitness of the modified organism or significantly altered its ecological niche. Fitness (the factor that reflects the rate at which a particular type of microorganism increases or decreases in number) might increase if a phenotypic change increased resistance to a noxious substance present in the environment or increased the ability of the organism to metabolize a substrate in the environment. It is often possible to determine the fitness of an organism through laboratory tests (Lenski, 1989), although field tests may be needed.

Considerable evidence exists that nonindigenous bacterial populations, including genetically altered strains, decline rapidly after they are introduced into soil or aquatic environments (Scanferlato, et al., 1989). This supports the well-documented fact that long-established microbial communities resist invasions by foreign organisms (Liang et al., 1982).

If persistence appears undesirable it may be avoided by choosing strains with reduced survival, reduced reproductive capacity, low resistance to a predictable change in the environment (such as seasonal heat or cold), or a tendency to lose the specific function of concern. Biological confinement also can be provided by suicide genes (Molin et al., 1987; Curtiss, 1988; OTA, 1988) or by incorporation of additional nutritional requirements. On the other hand, when persistence is desirable, it may be fostered to achieve the benefits for the intended function.

Persistence of the Genetic Modification

Gene transfer can affect persistence and may occur either to or from the introduced microorganism. Only a new genetic combination with higher fitness than the introduced or indigenous genotypes has any likelihood of persisting.

Conjugation, transduction, and transformation are mechanisms of exchange of both chromosomal and extrachromosomal DNA between bacteria (Freifelder, 1987; Lenski, 1987; Miller, 1988). Similarly, viruses can recombine when two or more types co-infect the same cell (Hershey and Rotman, 1949). Genetic material is exchanged most readily among clones of the same species, but exchange among groups that are more distantly related can occur as well (Roberts et al., 1977; Morese et al., 1986; Guerry and Colwell, 1977; Heinemann and Sprague, 1989).

Perhaps the most dramatic recombination mechanism is transformation, in which DNA from injured or dead cells (Kieft et al., 1987) or extruded by living cells (Borenstein and Ephrati-Elizur, 1969; Orrego et al., 1978) is taken up by living cells of the same or other species (Graham and Istock, 1979; Duncan et al., 1989).

We know very little about the exchange of genetic information among closely or distantly related microorganisms under natural conditions. The possibility that these exchanges occur, however infrequently, must be taken into consideration in any planned introduction. However, in the extensive trials carried out by the Monsanto Company and Clemson University in South Carolina, in which the *lacZY* genes were used as a marker for a genetically altered strain of *Pseudomonas fluorescens*, there was no evidence of exchange of the genetic marker with other soil bacteria (Drahos et al, 1988). Since those tests were limited both in time and location, it is evident that additional information must be generated on the frequency of gene exchange, particularly in managed ecosystems (Lindow et al., 1989).

Effective genetic transfer between distantly related microorganisms is far less frequent than between closely related microorganisms because of the improbability of each of a series of events: the insertion of DNA from a distantly related organism into the recipient organism; replication of the foreign DNA in the recipient; and selection favoring the new recombinant genotype. While we cannot assume that any given transfer is impossible in nature, a transfer of DNA among closely related organisms is far more probable than a transfer among distantly related ones. In principle, organisms manipulated in the laboratory to cross genetic barriers may be able in the field to transfer genes secondarily to other organisms; but in fact, even transfers between closely related microorganisms in nature are infrequent and difficult to document.

Not all genetic transfer is undesirable; in most instances it probably will not matter. Unless the recipient organism has a selective advantage, the genetic transfer will have little or no consequence; there will also be no consequence if an organism *has* a selective advantage, but is innocuous. Examples of genetic modifications that pose little or no risk due to genetic transfer include ice$^-$ (ice-nucleation deficient) *Pseudomonas syringae*, used to protect plants from frost damage, and plants in the rhizosphere containing the *lacZY* marker genes, used for monitoring their movement and persistence in soil.

PHENOTYPIC PROPERTIES AFFECTING THE ECOSYSTEM

If an organism persists in the environment and has the potential to spread, it is important to consider its phenotypic properties that relate to its role in the ecosystem. These properties include competitiveness, substrate utilization, environmental range, and host range (if the organism is a pathogen or a symbiont). These properties are also prime candidates for genetic modification, since many microbial applications depend on them. Examples include enhanced nitrogen fixation, altered host-range biocontrol agents, or beneficial soil bacteria that must out-compete indigenous microflora to be successful.

Competitiveness

Environments into which microorganisms are to be introduced are diverse and often support a complex indigenous microflora. As a result, competitive traits that are advantageous, such as a rapid utilization of abundant substrates, high maximum specific growth rate, and antibiotic production, may be necessary if an introduced microorganism is to colonize successfully in competition with native microorganisms.

Will the introduced microorganism persist or spread to other environments? It is difficult to envision a specialized microorganism faring unusually well in, for example, the plant rhizosphere because the soil microbial populations there are dense and incredibly diverse. However, when an organism is introduced into an environment in which the indigenous population is restricted in density and diversity, or in which the introduced microorganism has a monopoly on some resource (for example, a substrate difficult to metabolize), then the chances of proliferation to high density may be greater. However, spread probably will be limited if the resource is less abundant outside the target area.

It has been suggested that genetically modified microorganisms will be competitively disadvantaged, relative to their wild-type counterparts, because of burdens associated with carriage and expression of additional functions. Indeed there are many well-documented papers to support this suggestion (Lenski and Nguyen, 1988; Brill, 1985; Davis, 1987; Zund and Lebek, 1980; Lee and Edlin, 1985; Duval-Iflah et al., 1981). However, exceptions have been reported in which increased fitness has resulted from carriage of foreign genes in microorganisms (Hartl et al., 1983; Edlin et al., 1984; Bouma and Lenski, 1988). Most of these experiments have been performed in the laboratory and may not represent natural conditions. To date, how expression of additional functions affects fitness has not been clearly resolved.

It also has been suggested that an introduced microorganism might competitively exclude another potentially more valuable microorganism (Tiedje et al., 1989). An example cited is that *Bradyrhizobium japonicum* serogroup 123 is more competitive in some soils than in others and may exclude other more effective nitrogen fixers (Johnson et al., 1965; Moawad et al., 1984). As no basis other than field tests exists for judging field competition and no clear information is available on what governs competitive ability among the rhizobia, selection of inoculant strains has been made empirically. Strain 123 forms a symbiotic relationship with soybeans resulting in nitrogen fixation. In the infection process it outcompetes many strains in the field that fix nitrogen more vigorously than 123 under aseptic greenhouse conditions.

Substrate Utilization

It may be desirable either to expand or to restrict the range of substrates utilized for growth by an introduced microorganism. Expansion will have most appeal in applications for removal of stable compounds from polluted environments. As introduced toxin-degrading microorganisms proliferate, they and the indigenous microorganisms may co-exist owing to the removal of the toxic substance (Lenski and Hattingh, 1986). Genetic modifications also can be used to restrict substrate range, an appealing prospect for biological confinement of certain introduced microorganisms.

Involvement in Ecosystem Processes

Microorganisms play roles in ecosystem processes important to the sustained habitability of the planet, including major biogeochemical cycles: carbon (lignin and cellulose decomposition), nitrogen (nitrogen fixation, nitrification, denitrification), sulfur (sulfur oxidation and sulfate reduction), and the cycling of elements such as phosphorous, silicon, manganese, iron, and trace metals. Microorganisms also may produce or use gases that include CH_4, NO_2, H_2, and CO_2, and these gases may influence the earth's atmosphere. Field experiments should have no detectable influence on these processes but reasonable questions concerning potential adverse ecosystem effects should be evaluated prior to field testing.

Environmental Range

One of the classical criteria for choice of microorganisms for applied purposes is their ability to function under a variety of climatic and management conditions. The climatic range of a number of plant pathogens studied has been very restricted (for example, *Pseudomonas solanacearum*; Kelman, 1953). Another well-documented example covers the restricted range of root nodule bacteria. *Bradyrhizobium japonicum*, serogroup 123, dominates the root nodules of soybeans grown in the north-central region of the United States, despite a diversity of other strains of the same species found in the soils (Ham, 1980); however, serogroup 123 is rarely recovered from nodules of soybeans growing in the southeast states of the United States (Keyser et al., 1984). Similarly, *Rhizobium trifolii* TA1 has been a successful inoculant of clover species in eastern Australia but has been unsuccessful in western Australia (Parker et al., 1977).

It should be pointed out that non-indigenous or exotic saprophytic microorganisms have been introduced continually into soils in the United States for decades without documented cases of harmful responses. Examples of such introductions include those associated with shipment of plant propagation materials such as seedlings and bulbs that carry microorganisms.

Host Range

Host specificity is well documented in symbiotic and pathogenic associations between plants and microorganisms. Since relatively few microorganisms enter these relationships, they clearly are specialized. In recent years geneticists have identified not only the genes needed to establish symbioses or pathogenesis, but also genes that determine host range (Djordjevic et al., 1987; Keen and Staskawicz, 1988). From an agricultural perspective, it may be desirable either to restrict or to expand the host range of symbiotic microorganisms.

With certain plant pathogens, the incidence of disease relates in a complex manner to the population dynamics of the pathogen and its interaction with susceptible host cultivars (Rouse et al., 1985). Two strategies can be envisioned for biocontrol of plant disease: (1) a biocontrol agent is introduced to colonize and occupy the target habitat prior to the pathogen's appearance; or (2) the biocontrol agent competitively displaces the pathogen. In the first case, biocontrol agents with general features for rapid colonization of the leaf will have broad appeal. In the second, the biocontrol agent will usually need to be a specialized microorganism recognizing the same host specific-signals as the pathogen and out-competing the latter.

CHARACTERISTICS OF MICROBIAL PATHOGENS

One of the concerns often raised about genetic modification of microorganisms is the possibility that inadvertent acquisition or loss of one or a few functions might convert unrelated nonpathogenic organisms into potentially dangerous pathogens of humans, plants, or animals. However, pathogenicity is controlled by a large portion of the genome of prokaryotes. In certain plant-pathogenic bacteria, for example, many pathogenicity functions are carried on the chromosome, while others are carried on very large plasmids. Thus, the acquisition or loss of a single function cannot convert a nonpathogen into a pathogen. In some cases, changes in virulence (increased aggressiveness on specific hosts) may be increased by acquisition of the ability to produce a toxin or by loss of an incompatibility function, but the recipient organism must already possess all the wide variety of genes that will allow it to infect and colonize a particular host. Endowing rhizosphere bacteria with the ability to produce Bt toxin, as proposed by some biotechnology firms, does not represent conversion of a nonpathogen to a pathogen in the sense described above. Rather, a precise modification has been made that makes the bacterium toxic to certain insect larvae.

Background

Prokaryotes preceded eukaryotes during evolution. Thus, from the time higher plants and animals made their appearance on land, they have been surrounded by microorganisms. Parasitic or symbiotic relationships probably existed very early in evolution, but in spite of such long coevolution, relatively few microbial species have developed the capacity to parasitize plants or animals (Sequeira, 1984). For example, very few species of bacteria, representing only five major genera, parasitize plants. Among the gram-positive bacteria, which constitute a large portion of the soil microflora surrounding plant roots, there are no plant pathogens of significance with the exception of a few members of the Corynebacteriae. Even fewer species have developed a symbiotic relationship that allows them to multiply within plant cells. It is evident, therefore, that pathogenesis and symbiosis are highly complex relationships that developed, in relatively few instances, through eons of coevolution. It is equally evident that the evolution of a compatible relationship with a host cannot be recreated by simply inserting or deleting a few genes in a nonpathogen by recombinant DNA techniques.

Pathogenicity Versus Virulence

A distinction should be made at this point between genes that are involved in pathogenicity and those involved in virulence.

Pathogenicity is an attribute (the ability to cause disease) of an entire group, irrespective of the fact that particular strains or races may not be pathogenic to a given host. Virulence is the relative ability of an individual strain or race to cause disease under defined conditions (Federation of British Plant Pathologists, 1973). Thus, virulence is a quantitative variable. While it may be appropriate to refer to a particular strain as nonvirulent, the term nonpathogenic may not be appropriate unless a wide range of hosts has been tested.

Genes That Confer Pathogenicity

Pathogenic microorganisms must possess two general types of genes: (1) those that are important for basic compatibility with the host; and, (2) those that are specific for virulence on particular hosts. To the first category belong housekeeping genes that control general metabolism and that allow the pathogen to grow on the nutrients available in the host. To this same group belong those genes that allow the pathogen to degrade, detoxify, or bypass preformed or induced substances that provide resistance for the host. Thus, one can assume a priori that a pathogen of a particular host must be able to utilize the nutrients available at the site of infection and must be insensitive to substances that inactivate or destroy the vast majority of the microorganisms that invade the host. The marked specificity of certain strains of bacteria that colonize leaf surfaces, for example, indicates a narrow adaptation to the nutritional and toxic components of the host environment. Also, these bacteria must be able to withstand intense solar radiation and extended periods of desiccation. Once inside the leaf, those that are pathogens now produce enzymes that degrade plant cell walls or toxins that may be very specific in terms of the cellular targets affected in certain tissues of particular hosts. These virulence functions are controlled by genes that belong to the second category.

Gene-for-Gene Interactions

Interactions between pathogens and their hosts are often under the control of apparently gene-for-gene systems (Flor, 1956; Barrett, 1985). Such systems are common in plant-pathogen relationships, where host resistance depends on single, dominant genes that are superimposed on other, more general mechanisms of defense. The gene-for-gene concept accounts for the fact that for every major resistance gene in the host there is a corresponding gene for avirulence in the pathogen. For many pathogens, there is a great deal of evidence that avirulence, rather than virulence, is the positive function. Important consequences follow from this seemingly contradictory statement. First, incompatibility of a pathogen on a particular host may result from the interaction of products of an avirulence gene of the pathogen and the corresponding resistance gene in the host (Ellingboe, 1982). Second, mutations that eliminate the avirulence gene product may increase the pathogenic potential of an organism (Staskawicz et al., 1988; Mellano and Cooksey, 1988). Ultraviolet treatment of plant parasitic fungi, for example, has long been known to yield mutants that have a wider host range than that of the parent line (Flor, 1958).

There are, of course, many examples of positive-control functions that also specify virulence of pathogens. Of particular interest are the host-specific toxins produced by some pathogenic fungi that have highly specific targets in particular varieties of the plant host. A particular receptor for the toxin appears to confer susceptibility. The toxin from *Helminthosporium maydis*, for example, affects only the mitochondria of hybrid corn lines carrying Texas male sterile cytoplasm (Yoder, 1980). Other species of *Helminthosporium* produce toxins that affect varieties of oats and sugar cane. Artificial hybrids among species of *Helminthosporium* may exhibit combined virulence to each of the hosts affected by the parental strains, as specified by the toxins they produce.

GENERAL PHENOTYPIC CHARACTERISTICS OF PATHOGENS

One of the primary characteristics of pathogens is their ability to colonize the host. For many microorganisms, the ability to attach to the host surface, or to specific host cells in particular tissues, is the first step in colonization (Costerton et al., 1981). Microorganisms have developed highly specialized structures (extracellular polysaccharides, fimbriae, appressoria) that provide adhesion to specific host surfaces. Once attached, colonization depends not only on the ability of the organism to utilize nutrients at the site, but also on the ability to compete with other organisms for those same nutrients. A high degree of competitiveness, therefore, characterizes a successful pathogen. Saprophytes share many of these properties with pathogens, but they may be unable to colonize particular ecological niches within a host because of their inability to resist a wide range of host defense systems.

A series of nonspecific and specific mechanisms for resistance present a hostile environment that limits the growth of the vast majority of microorganisms that gain access to an uncompromised host (Falcone et al., 1984; Mims, 1982). With the exception of commensal species, only those organisms that have an array of virulence factors are capable of growing in animal or plant host environments. Extracellular pathogens of animals, for example, owe their virulence to their ability to resist or inhibit phagocytosis. The production of capsular polysaccharides is important in this regard. Intracellular pathogens, on the other hand, are highly adapted to ingestion by phagocytes and even use phagocytes as a means for dispersal within the host. There are evident analogies between plant and animal pathogens in this regard. Whether intracellular or extracellular, the ability to resist, inhibit, or degrade host defense compounds when colonizing the host is therefore one of the key properties of pathogens. Some plant pathogens, for example, are capable of chemically degrading phytoalexins, antimicrobial compounds produced by plants upon challenge by potential pathogens. Others are insensitive to phytoalexins from particular hosts. Yet others grow in the host in such a way that they effectively bypass the tissues where phytoalexins accumulate.

Pathogens can establish themselves in the host because of their ability to produce a large series of compounds, including hydrolytic and proteolytic enzymes, toxins, polysaccharides, and growth regulators that affect the host in ways that ultimately favor multiplication or transmission of the pathogen. The enzymes that destroy cell membranes or cell walls, the toxins that affect the normal metabolism of the cell, and the growth regulators that influence the ways host cells grow all have an impact in determining pathogenesis. As a result, the host is damaged much more than would be expected from the strictly energetic drain on its metabolism.

Pathogens must also have an effective mechanism for spreading to new hosts. Reproductive structures must be formed that can persist in unfavorable environments or to reach new hosts before they perish. Most pathogenic organisms depend on their hosts or on vectors for survival, but many are capable of surviving for long periods in soil or water (Brubaker, 1985). The chemotactic property of microorganisms often assumes importance in this respect since many animal pathogens must move through intercellular spaces to reach target tissues.

Acquisition of Virulence

It is evident from this discussion that pathogenicity depends on an impressive array of different characteristics that relatively few microorganisms have acquired through extended coevolution with particular hosts. It is unlikely, therefore, that minor genetic modifications can convert a nonpathogen into a pathogen. It also is clear that factors for virulence are not exchanged indiscriminately among microbial populations. Although certain unique properties of pathogens are encoded by plasmids and may be transferable under certain conditions, only a limited portion of the whole array of genes required for virulence would be acquired by the recipient organism. These plasmids, therefore, can confer virulence only to related strains that are already highly adapted to particular ecological niches on or in the host. For example, most of the genes necessary for tumorigenicity of *Agrobacterium tumefaciens*, the crown gall pathogen, are located in a plasmid (Ti). Acquisition of the Ti plasmid by strains of *Agrobacterium radiobacter*, a nonpathogenic soil inhabitant, automatically converts them into the pathogen *A. tumefaciens.* There is a great deal of evidence, however, that the two species are essentially identical except for the presence of the plasmid (Nester and Kosuge, 1981).

Similarly, *E. coli* is a normal component of human and some animal intestinal microflora, but certain strains are pathogenic and can cause different types of diarrhea or urinary tract infections. Certain commensal strains of *E. coli* become capable of causing disease upon acquisition of plasmids conferring enterotoxicity and adhesiveness (McConnell et al., 1981). These pathogenic strains, however, often have short persistence times, presumably because they are less fit than strains lacking these elements (Duval-Iflah et al., 1981).

In these examples, virulence is increased only when factors from very closely related species are acquired. When changes are effected between unrelated species, pathogenicity is rarely acquired. For example, transfer of pectin lyase genes from *Erwinia chrysanthemi*, a potato soft-rot pathogen, to *E. coli* does not convert the latter into a pathogen, even though it is now capable of rotting potato tuber slices in the laboratory (Collmer and Keen, 1986).

The overall conclusion of this discussion is that, as a rule, an unrelated microorganism does not become a pathogen merely because it has received a portion of the DNA of a pathogenic species.

SUMMARY POINTS

1. From the standpoint of assessing potential effects of unfamiliar application of microorganisms in managed ecosystems, the most relevant phenotypic properties are those that relate to the persistence of the microorganism (and its genetic modification) and those properties that may have adverse effects. Such assessments are not necessary, however, if a proposed microbial application is familiar and has a safe history of usage in the environment.

2. Key phenotypic properties include the fitness of a genetically modified microorganism relative to its unmodified counterpart; the potential for gene transfer between the introduced microorganism and the indigenous microflora; the physiological tolerances of the introduced microorganism; the competitiveness of the introduced microorganism; the range of substrates available to the introduced microorganism; and, if applicable, the pathogenicity, virulence, and host range of the introduced microorganism.

3. Persistence of genetic modifications may occur in either of two ways. The introduced microorganism itself may survive and propagate to form a self-sustaining population. Alternatively, the modification may persist in new genetic combinations resulting from gene transfer. The potential effects of undesired persistence can be avoided by the use of appropriate strategies for biological confinement of the introduced microorganism and its genetic material.

4. Microorganisms intended for environmental introduction may be modified for either highly specialized or more generalized phenotypic properties, such as competitiveness, substrate utilization, physiological tolerance, and host range (either pathogen or symbiont). Persistence after introduction is most likely for microorganisms having generalized properties and for microorganisms having properties that allow exploitation of previously unfilled ecological niches.

5. Pathogenicity is controlled by a large portion of the genome of microorganisms. This is so because a pathogen must be able to cross a large number of potential barriers to be successful. The

successful pathogen must be able to attach to appropriate sites on or in the host, to compete for nutrients within the host, to resist various host defenses, to survive and persist when the host is not present, and to be effectively transmitted to new hosts. Therefore, acquisition of a small number of genes from a pathogenic source cannot convert an unrelated, nonpathogenic microorganism into a pathogen.

6. However, if the recipient is closely related to the pathogenic source, or if the recipient is itself a pathogen, increased virulence for particular hosts may result. In these special cases, the recipient already contains a large complement of genes related to pathogenicity.

10
Properties of the Environment Relevant to the Introduction of Genetically Modified Microorganisms

Successful establishment of a specific population depends on two components: the organism and the environment. Although most discussion of genetically modified microorganisms has focused on the properties of the organisms, properties of the environment are equally important.

Microbial environments often are less well understood than those of higher organisms because of the difficulties in defining, measuring, and controlling various physical and chemical details of the microenvironment important in establishing introduced microorganisms. More studies of microbial interactions with the environment are needed. However, we have considerable knowledge about certain microorganisms, such as *Rhizobium* and mycorrhizal fungi that provide important insights into the interactions in question.

TYPES OF ENVIRONMENTS

Environments that are to receive introduced microorganisms may vary considerably with respect to their biological, chemical, and geological properties, and these properties may vary with physicochemical changes. Hence, our ability to establish a level of certainty about the risks and benefits varies with our knowledge and experience with the particular site where the microorganisms are to be introduced. Scientists have extensive experience with major agricultural soil-crop environments, much less experience with tropical forests, and little experience with the open ocean. Predictability about the fate and effects of introduced microorganisms increases with experience.

The ease and reliability with which the introduced microorganism can be confined also depend on the environment into which it is introduced. Microorganisms introduced onto surface soil not subject to excessive wind and rainfall are obviously more easily confined to the test site than are microorganisms added to a site subject to flooding and excessive wind erosion.

Special features of the environment may be important in deciding whether an introduction is advisable. Most introductions are intended for sites that are far from pristine. At hazardous waste sites or in streams made acid by mine drainage, a microbial introduction will promise far more benefit than risk. Microorganisms designed for removal of toxic pollutants would be unlikely to flourish outside the site of introduction.

HABITABILITY OF ENVIRONMENTS

Four characteristics of the environment that determine habitability of an area for introduced microorganisms are (1) nutrient status, (2) toxic chemicals and metabolites, (3) physicochemical factors, and (4) biological factors.

Nutrient Status

Energy supply often limits the growth of microbial populations. Organic compounds represent the major energy source for most genetically modified microorganisms currently being studied. Light or reduced inorganic compounds also can supply energy for some microorganisms. Although population density and community diversity usually parallel the organic carbon concentration in a habitat, the competition for these carbon substrates and the diversity of substances are also important. Whereas energy usually limits heterotrophic populations, inorganic nutrients may limit others. For example, algal blooms are typically limited by phosphate in freshwater environments and by nitrogen in the open ocean. Carbon and nutrient resources are limited in nature, and the growth of introduced microorganisms cannot exceed these resource limits.

Toxic Chemicals and Metabolites

Extreme concentrations of heavy metals, acids, and organic pollutants can be toxic to microbial cells through their effects on metabolic processes. Thus, toxic compounds may decrease microbial population density and limit community diversity to microorganisms resistant to elevated concentrations of heavy metals and organic pollutants. Bacterial resistance to heavy metal toxicity can indicate whether certain metals are or have been present in a given environment (Olson and Thornton, 1982; Olson and Barkay, 1986; Zelibor et al., 1987). Specialized microorganisms often successfully colonize these stressed habitats, but may be relatively less competitive in nonstressed environments (Konings and Veldkamp, 1980).

Physicochemical Factors

Environmental chemical variables (pH, oxidation/reduction potential, nutrients, toxicants, salinity) and physical variables (light, surfaces, temperature) influence the diversity of microbial communities (Stotzky and Babich, 1986). Environmental factors such as moisture, temperature, and oxygen can vary over wide ranges.

Biological Factors

Microbial predators, parasites, symbionts, and competitors contribute significantly to microbial community structure. Experience from microbial introductions into soil and water environments typically has shown that it is difficult to establish introduced populations at densities sufficient to achieve the desired effect, such as nitrogen fixation in rice paddies (Reddy and Roger, 1988) and biocontrol of pests (Bahme and Schroth, 1987). The lack of success of some microbial introductions may result from competition with the indigenous community or introduction into unfavorable habitats.

DISPERSAL

An important issue for the environmental introduction of microorganisms is the extent to which the microorganism and its progeny are dispersed from the application site. Dispersal provides a route of entry for microorganisms to new habitats. Prior experience has revealed no problems that have arisen because introduced microorganisms became established beyond their area of introduction, with the exception of pathogens. Even with plant pathogens, carefully designed field tests are routinely performed which result in no or negligible damage to neighboring crops (Tolin and Vidaver, 1989). Exceptions have occurred, usually in the field testing or introduction of plant pathogens that are not indigenous to the area. Microorganisms can be dispersed from terrestrial field plots during their application, or by leaching through soil, runoff in surface water or on soil particles, dissemination by wind (dust particles), and transport from the plot by animals, humans, and field machinery. Soil generally filters microorganisms effectively, and the motility of soil microorganisms does not support extensive movement.

Most microorganisms in soil are firmly attached to soil particles, so that any movement of the soil particle moves the attached organism as well (Faust, 1982). Attachment often aids microbial survival, because particles can protect organisms from ultraviolet radiation during aerial transport (Stetzenbach, 1989) and from predation in soil and water (Roper and Marshall, 1974).

Recently introduced organisms may be less likely to attach to the soil, and they may move by saturated flow (Rake et al., 1978; Smith et al., 1985). Laboratory studies with sieved and repacked soil cores tend to underestimate microbial movement by leaching compared with that found in undisturbed soil cores, because natural soil structure has more macropores and connecting channels that reduce its effectiveness as a filter.

Groundwater and surface water environments may furnish similar habitats and harbor similar heterotrophic microorganisms. Digestive tracts of insects, birds, and other animals provide a habitat quite different from soil, but they may support soil organisms and disperse them with fecal material (Reyes and Tiedje, 1976).

If the major dispersal mechanisms are known, dispersal can often be effectively evaluated and controlled. Selection of level sites for field tests and construction of terraces should virtually eliminate surface runoff. Selection of a site distant from groundwater and application of the organisms to minimize individual cell movement will reduce leaching to groundwater. Fencing and netting can be used to control animals. Controlling insect dispersal of the introduced microorganism may be difficult, but such dispersal often can be minimized by the choice of suitable plants and through the use of appropriate insecticides.

Small-scale tests in estuarine, marine, and other aquatic environments pose special problems of confinement. The use of membrane chambers offers confinement (McFeters and Stuart, 1972; Mach and Grimes, 1982; Grimes and Colwell, 1986), but the ability of marine microorganisms to shrink during starvation (Morita, 1982) or to enter a dormant state (R. R. Colwell et al., 1985; Grimes et al., 1986; Roszak and Colwell, 1987) requires that membrane chambers have sufficiently small pore size to retain cells but still permit nutrients to enter the chamber. "Bubble" containment devices placed in the aquatic environment are not reliable protection against dispersal (Grice and Reeve, 1982), and tests of genetically modified microorganisms that might be performed using them should be carefully evaluated. Laboratory microcosms and mesocosms are the most suitable compromise (Pritchard and Bourquin, 1984; Cripe and Pritchard, 1989). In estuarine, coastal, and open ocean systems, the effects of wind, tides, and currents, as well as dispersal by fish, birds, aquatic plants, and other organisms, must be considered.

SUITABILITY OF MICROCOSMS FOR TESTING OF MICROBIAL INTRODUCTIONS

It is widely accepted that aquatic and terrestrial laboratory microcosms are useful for examining the fate and effects of introduced microorganisms as well as their survival and persistence in specific environments. The definition of a microcosm which can be adapted for purposes of discussion here is ". . . an intact, minimally disturbed piece of an ecosystem brought into the laboratory for study" (Cripe and Pritchard, 1989, p. 1). Thus, a microcosm can be used to relate laboratory data to the site where the environmental samples were taken (Greenberg et al., 1988) as it is a site- and system-specific construct. In principle, the microcosm is an intact piece of the field that behaves ecologically like its counterpart in the actual field (Pritchard and Bourquin, 1984).

A variety of laboratory test systems have been designed to model the environment. These include synthetic communities, with well-characterized organisms placed in sterile media under defined environmental conditions. In other systems, natural samples may be incubated over long periods such that a unique and sustaining ecosystem evolves (Greenberg et al., 1988). Results of several studies attest to the value of studying ecological processes in microcosms and extrapolating the information to ecosystems in nature (Livingston et al., 1985; Diaz et al., 1987).

Most use of microcosms has been applied to the study of toxic substances in the environment (Gillett and Witt, 1979; Giesy, 1980; Hammonds, 1981; Cairns et al., 1981). Synthetic communities have proven useful in studying the fate of xenobiotic compounds (foreign chemicals) in aquatic systems (Isensee and Tayaputch, 1986; Metcalf et al., 1971) as well as their effects on biological communities (Crow and Taub, 1979; Leffler, 1984). Artificial communities may be limited, however, in that they lack complex population structures and may not function like those occurring in natural ecosystems (Cripe and Pritchard, 1989).

Some of the ecological processes quantified in microcosms include nutrient leaching (Van Voris et al., 1980, 1983), nutrient cycling (Harte et al., 1980), predator-prey interaction (Gillett et al., 1983), primary production (Harte et al., 1980), and microbial respiration (Lighthart et al., 1982; Taub and Crow, 1980).

Model ecosystems, although often criticized as ecologically simplistic, have been heavily used in assessments of pesticides and toxic substances, as in the constructed "farm-pond" system of Metcalf et al. (1971). If it had been used decades earlier, such a model might well have suggested possible adverse ecological consequences of the use of chlorinated hydrocarbon pesticides (Gillett et al., 1985). Model ecosystems also may be useful for monitoring the colonization and persistence of genetically modified microorganisms. Single-species tests, synthetic communities, and microcosms provide three preliminary field-trial assessments of ecological effects. Microcosms, if operated in a manner simulating the field site, sometimes may be used as surrogates for field research, with reduced effort and cost (Cripe and Pritchard, 1989). If the measurements taken to analyze a microcosm are the same as those used in the field, microcosms can be used to establish the sensitivity and appropriateness of analytical methods pertinent to field tests.

Microcosm studies are most useful in situations when their performance is needed to clarify questions about unfamiliar introductions. They are not necessary in cases where experience and scientific inference provide enough information to permit scientists to understand and be familiar with the intended introduction. Furthermore, the microcosm is a site- and system-specific model; specific processes may be of greater or lesser importance depending on the site and ecological system (Cripe and Pritchard, 1989). Effects that are strongly scale-dependent may be overlooked in microcosm studies.

In instances when such studies are deemed to be appropriate, questions of biological containment and certain environmental effects can be addressed. These include persistence in a given environment, transfer of genetic material to other organisms, population density and community structure, changes in heterotrophic activity, and nutrient cycling (Cripe and Pritchard, 1989). Gillett et al. (1985) concluded that microcosm technology should receive a high priority for assessment of both hazard and exposure.

SCALE AND FREQUENCY OF INTRODUCTION

In this report we cover small-scale field tests and not large-scale introductions of organisms. The report sponsored by the Ecological Society of America (Tiedje et al., 1989), documents the importance of scale and encourages the use of small-scale field tests, when appropriate, to evaluate the potential for larger scale environmental effects.

POTENTIAL EFFECTIVENESS OF MONITORING

Many planned introductions of genetically modified microorganisms should include appropriate methodology for monitoring the released microorganisms in and around the test site. Monitoring is important for several reasons: (1) understanding the basis for the organism's effectiveness, (2) detecting any unexpected spread, and (3) building a data base on survival, spread, genetic stability, and ecological effects of genetically modified microorganisms in nature.

Lack of efficient recovery of the microorganisms and insensitive assays are often obstacles to monitoring microbial populations introduced into the environment. The classical plate-count method or similar culture methods are still the mainstays of monitoring protocols. Stressed or dormant organisms may not be recovered by culturing (Roszak and Colwell, 1987). The most common markers used for tracking microbial populations are antibiotic resistances; a typical lower limit for detection is in the range of 10^3 organisms per gram of soil. At this level of detection, for a 1-hectare field site, to a depth of 10 centimeters, about 10^{12} microorganisms could survive and yet be undetectable. Thus, it often has little meaning to argue whether an introduced organism dies out completely. Rather, it is more important to focus on whether a residual population can multiply under the environmental conditions expected at the site of testing.

More sensitive and less costly monitoring methods are needed. Reliance on antibiotic-resistance as a selective marker has been extensive, but spreading antibiotic-resistant strains should be discouraged. The advantages and disadvantages of selective culturing as well as antibody and nucleic acid hybridization methods for monitoring introduced transgenic organisms in the environment have been summarized elsewhere (Tiedje, 1987; R. K. Colwell et al., 1988).

The nucleic acid methods offer the most specificity, require no prior culturing, have high potential sensitivity, and thus, recently, have received the most attention. These methods include (1) detection by DNA-DNA hybridization of unique ribosomal sequences in total DNA extracted from communities (Attwood et al., 1988); (2) detection by microscopy of cells containing a DNA fluor hybridized to unique ribosomal RNA sequences in cells (Giovianni et al., 1988); (3) detection by DNA-DNA hybridization of cloned genes in the total DNA extracted from soil and seawater communities (Holben et al., 1988; R. R. Colwell et al., 1988; Somerville et al., 1989); (4) detection of unique but native sequences in the total DNA of communities (Steffan et al., 1988); and (5) detection by DNA hybridization after DNA amplification by polymerase chain reaction to improve sensitivity (Steffan and Atlas, 1988). Nucleic acid methods have been combined with culturing methods, such as the most probable number method (Fredrickson et al., 1988) and plate counts (Ogram and Sayler, 1988) to improve specificity and sensitivity when culturing is not a limitation.

Polyclonal antibody methods have long been used in microbial ecology (Bohlool and Schmidt, 1980), and recently monoclonal methods have been used to improve specificity (DeMaagd et al., 1989; Wright et al., 1986). Although these methods are excellent when used to study the ecology of the indigenous community, they may not always distinguish between the introduced organism and its indigenous close relatives unless combined with nucleic acid methodology to track the specific genetic material.

MITIGATION

Microbial habitats vary in the ease and effectiveness with which unwanted effects from introduced organisms can be mitigated. In most natural environments it is possible to reduce populations but difficult to eliminate an introduced microorganism completely if it becomes established. Control methods have been summarized by Vidaver and Stotzky (1989). For terrestrial environments, the methods include fumigation, sealing the soil surface, or otherwise making the introduced microorganism's environment unfavorable for survival.

Some fumigants are selective for fungi and do not kill bacteria, but methyl bromide has proven effective in controlling plant pathogenic bacteria as well as fungi. Its effectiveness depends on soil texture, moisture, and the depth at which the organisms are located; proper use of methyl bromide under a tarpaulin should control most introduced fungi and bacteria. Many antibiotics and nonvolatile organic biocides are ineffective in soil because they are readily bound to soil material and cannot be mixed effectively throughout the soil volume.

The environment may be made inhospitable to the introduced organisms by flooding the soil to create anaerobic conditions, altering the pH by adding lime or sulphur, or destroying the plant vegetation by burning or other means.

Effort should put into characterizing and preventing risk, so that mitigation plans are only a secondary means of environmental protection. Unfortunately, few if any of these methods are applicable to aquatic ecosystems, as estuarine and marine systems are driven by tides and currents, and lakes are subject to substantial mixing; material cannot be confined after its introduction.

SUMMARY POINTS

1. The persistence and effects of an introduced microorganism depend on features of the environment as well as on phenotypic properties of the organism. Some environments are better understood than others; for example, we know more about agricultural fields than natural ecosystems.

2. The ease and reliability with which a particular introduced microorganism can be confined are important considerations in choosing a target environment for a field test. The potential for dispersal of introduced organisms, their progeny, or their genes by exchange with indigenous organisms must be considered.

3. Environmental features that will affect the likelihood of persistence of an introduced microorganism include nutrient availability; physicochemical factors such as pH, temperature, and inhibitory chemicals; and biological factors such as competitors and predators.

4. Although documented examples of introduced microorganisms that have measurably or adversely altered ecosystem processes are not available, unfamiliar microorganisms should be studied carefully first in the laboratory and then in small-scale field tests before they are introduced on a broader scale.

5. Microcosms are minimally disturbed "pieces" of natural ecosystems that are brought into the laboratory, and their use in appropriate cases may provide useful information for evaluating the survival and impact of proposed microbial field tests.

6. Small-scale field research is an important step in the investigation of properties of a particular microorganism intended eventually for environmental application. Hence, field research should be encouraged after appropriate investigations have been conducted in physically confined settings and after appropriate methods have been considered for monitoring and controlling the introduced microorganism.

11
Conclusions and Recommendations: Microorganisms

Mankind has a long history of using microorganisms in food processing, agriculture, waste treatment, and in other beneficial applications. New molecular methods for genetically modifying microorganisms will expand the range of beneficial applications, for example, in control of plant disease and in biodegradation of toxic pollutants.

In many respects, molecular methods resemble the classical methods for modifying particular strains of microorganisms, but many of the new methods have two features that make them even more useful than the classical methods. *Precision* allows scientists to make genetic modifications in microbial strains that can be characterized more fully, in some cases to the level of the DNA sequence. This reduces the degree of uncertainty associated with any intended application. The new methods have greater *power* because they enable scientists to isolate genes and transfer them across natural barriers.

The power of these new techniques creates the opportunity for new applications of microorganisms. Despite some initial concerns over the use of recombinant methods in laboratory research, it is now clear that these methods in themselves are not intrinsically dangerous.

The next step after laboratory experimentation is to test modified microorganisms in the field, and establishing a scientifically based framework for decisions on field testing has been a primary purpose in this report. No adverse effects of introductions have been seen and an extensive body of information documents safe introductions of some microorganisms, such as the rhizobia, mycorrhizal fungi, baculoviruses, *Bacillus thuringiensis*, and *Agrobacterium radiobacter*. However, less is known about field tests of microorganisms than of plants. Thus, for unfamiliar applications, it is prudent to prepare for the control of the introduced microorganisms.

Questions concerning the effects of an introduced microorganism arise whenever the intended introduction differs substantially from those with an established record of safety. Such questions as unintended persistence and possible adverse effects should be addressed scientifically, and as the scientific community continues to accumulate information regarding the safety or risk of environmental applications of microorganisms in field tests, levels of oversight can be tuned to the needs of particular situations.

In the recommendations that follow, a framework has been developed as a basis for a workable and scientifically based evaluation of the safety of microorganisms intended for field testing. This framework has been developed from consideration of three criteria: (1) familiarity with the history of introductions similar to the proposed introduction (Chapter 7), (2) control over persistence and spread of the introduced microorganism as well as over exchange of genetic material with the indigenous microflora (Chapters 8 through 10), and (3) environmental effects, including potential adverse effects associated with the introduction (Chapters 9 and 10).

The framework does not distinguish between classical and molecular methods of genetic manipulation, nor between modified and unmodified genotypes. The framework is product- rather than process-oriented, focusing on the properties of the microorganism rather than on the methods by which it is obtained. Knowledge of the methods used may nonetheless yield useful information concerning the precision of genetic characterization of the microorganism, which in turn may be relevant for assessment of its similarity to previous applications, persistence, and possible effects after introduction.

The framework has not focused on other variables, often suggested as criteria for oversight, because they convey relatively less scientifically useful information for assessments: the sources of genes, whether recombinants are intra- or intergeneric, and whether coding or noncoding regions of the genome have been modified. The necessity of using, whenever possible, simple and readily identifiable criteria for oversight is recognized.

Terms such as "uncertainty," "sufficient," and "significant" are used in the framework without precisely defining their quantitative limits. Any specific numerical values assigned would be arbitrary and subject to disagreement, as some underlying variables may be difficult to quantify precisely. In the final analysis, assignment of risk categories must include a rational examination of the relevant scientific knowledge for each introduction.

In the framework, assessments of potential risks arising from the introduction of microorganisms into the environment are made according to the three major criteria of familiarity, control, and effects. Upon evaluation of these three criteria, a proposed introduction can be field-tested according to established practice or it can be assigned to one of three levels of concern: low, moderate, or high uncertainty (Fig. 11-1). The framework is inherently flexible, allowing an application to be reassigned to a different category as additional scientific information is obtained that is relevant to any of the three criteria.

Small-scale field tests can proceed according to established practice if the microorganism used, its intended function, and the target environment are all sufficiently similar to prior introductions that have a safe history of use (Fig. 11-2). *Rhizobium* used for enhancement of nitrogen fixation in leguminous crops provides a familiar example.

If an introduction does not satisfy the familiarity criteria, it is evaluated with respect to both our ability to control the microorganism's persistence and dissemination and the microorganism's potential for significant adverse effects (Fig. 11-1). For example, *Rhizobium* modified to encode an insecticidal toxin would not be a familiar introduction, even though it might well prove to be safe. An introduction is considered to be in the low-uncertainty category if it satisfies appropriate criteria with respect to both controllability and low potential to result in adverse effects. An introduction is considered to be in the moderate-uncertainty category if it satisfies criteria for either controllability or potential effects, but not both. An introduction is considered to be in the high-uncertainty category if it satisfies neither the control nor the effects criterion (Fig. 11-1). The high uncertainty status implies that potential adverse effects exist and are coupled with potential inability to control the microorganism, and hence its potential effects.

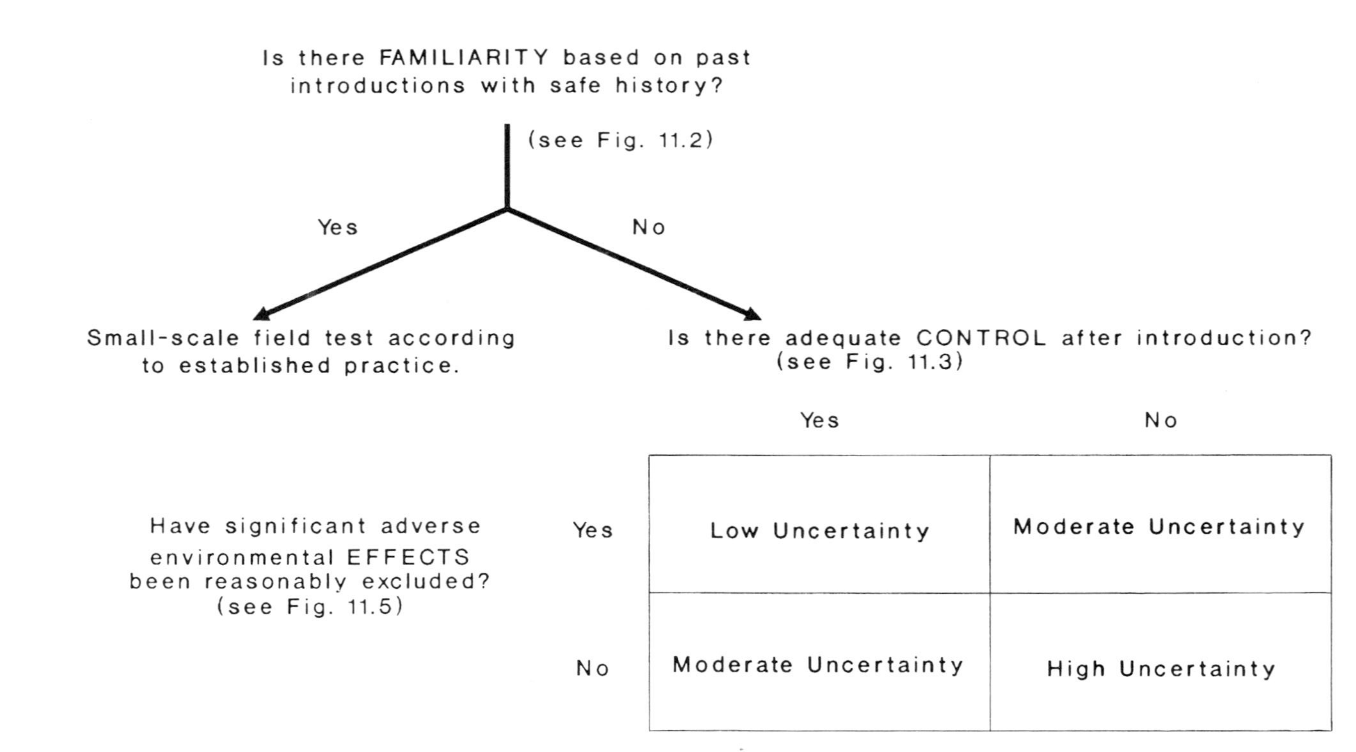

FIGURE 11.1 Framework to assess field testing of genetically modified microorganisms.

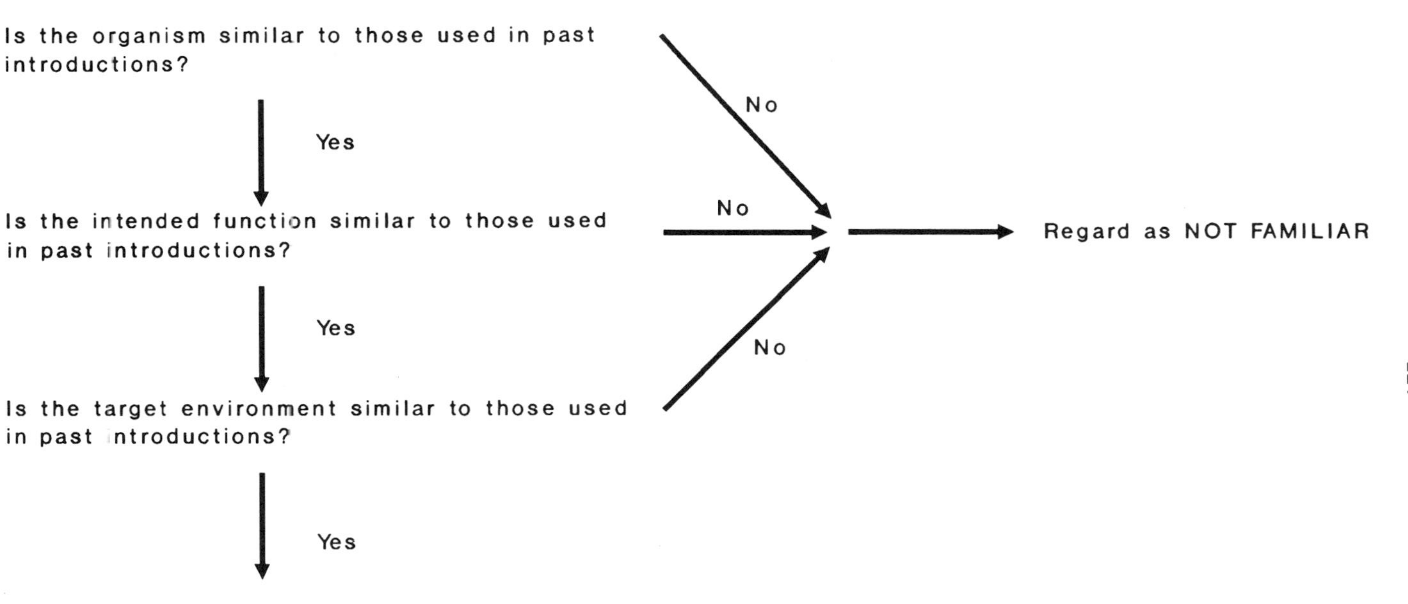

FIGURE 11.2 Familiarity.

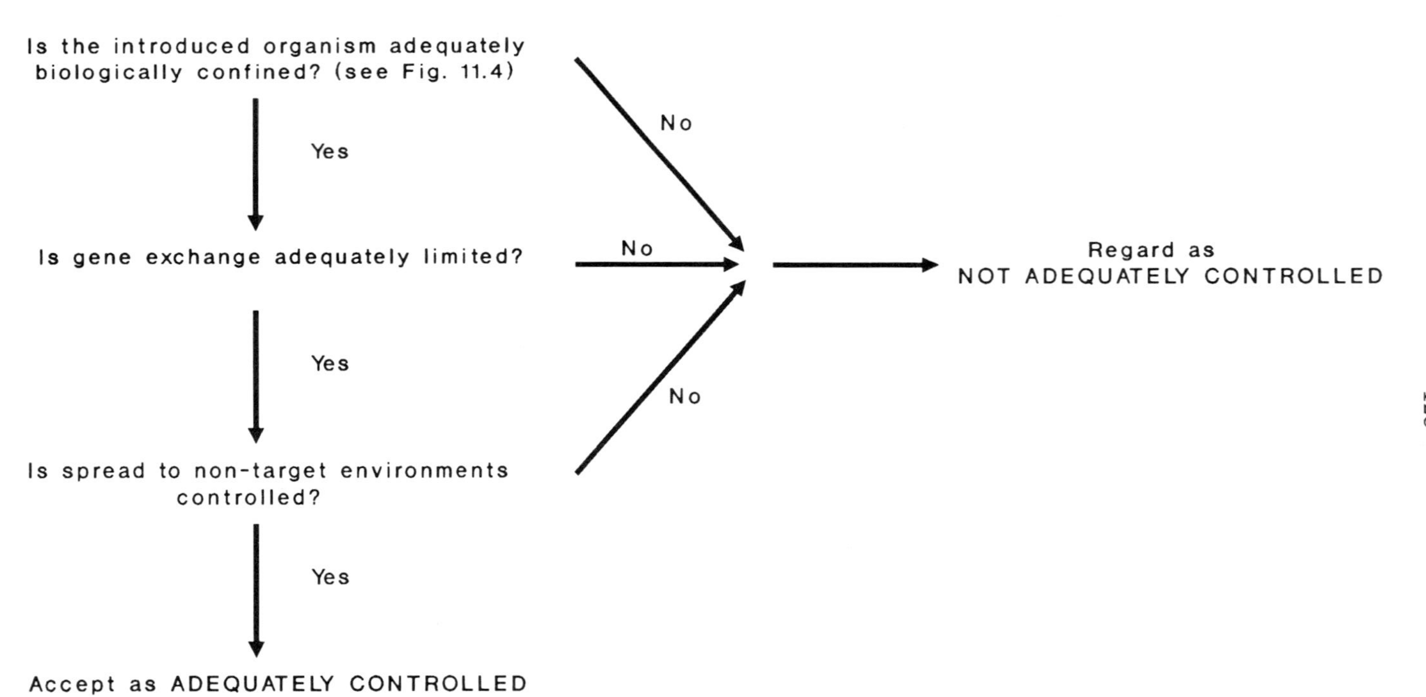

FIGURE 11.3 Control. Appropriate questions for specific applications to be added by users of the framework.

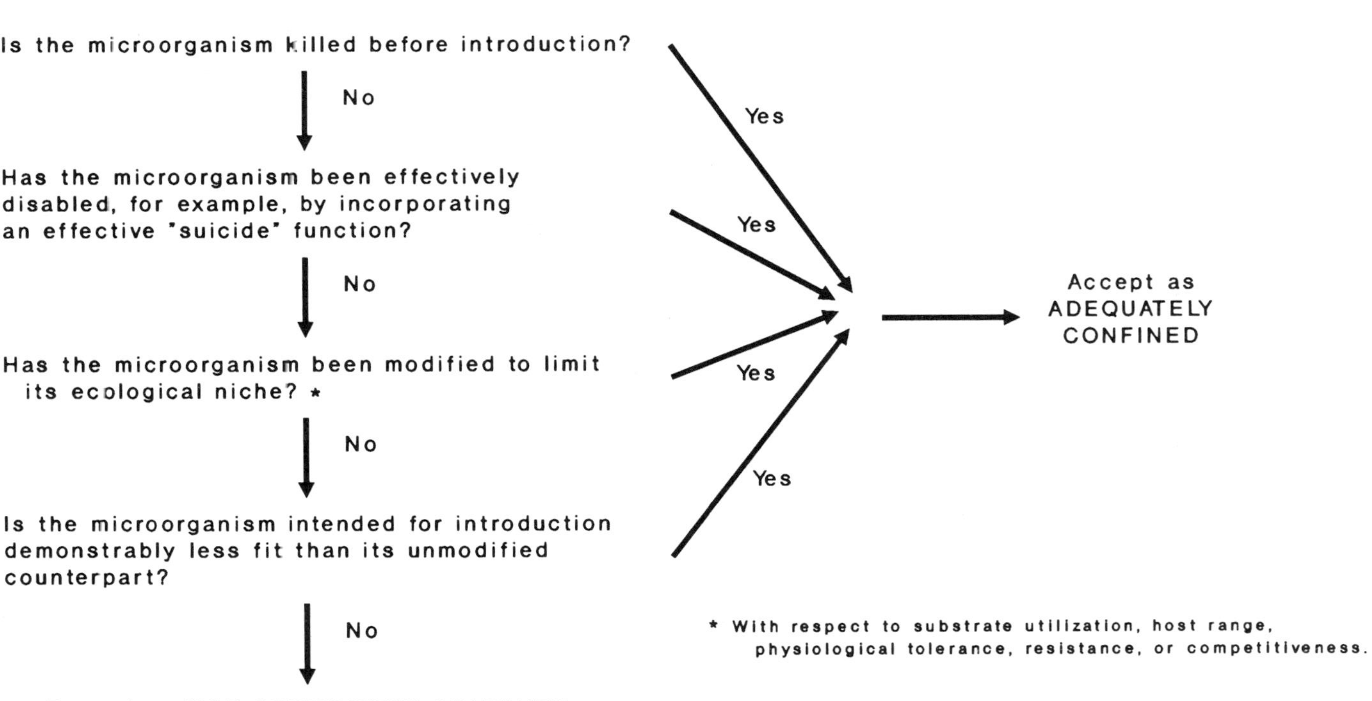

FIGURE 11.4 Biological confinement. Appropriate questions for specific applications to be added by users of the framework.

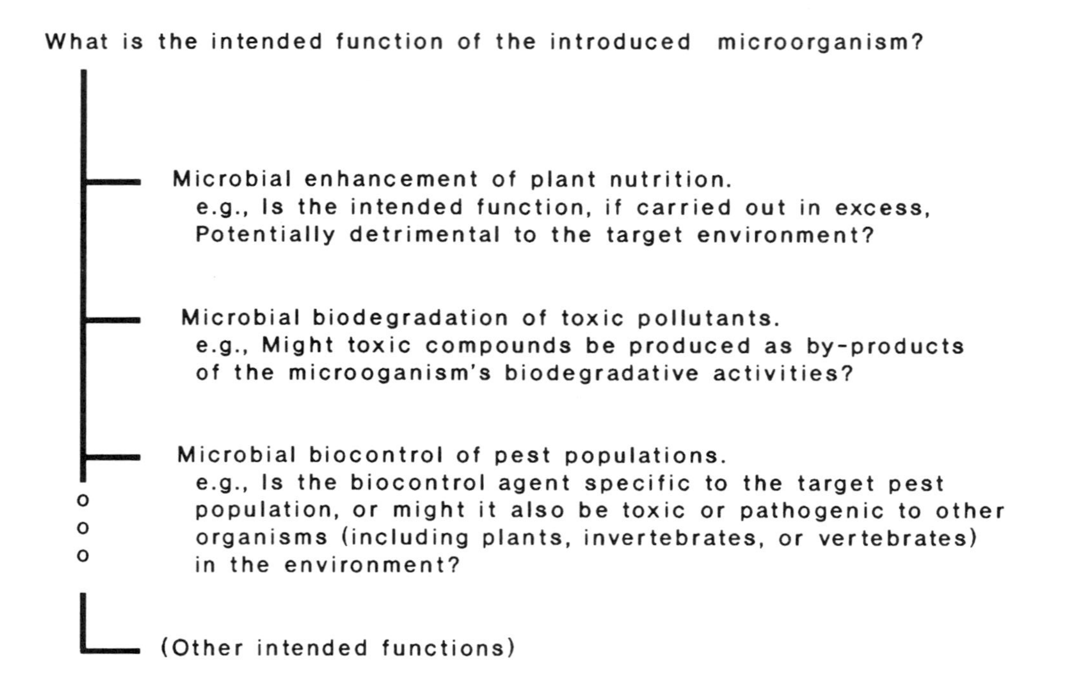

FIGURE 11.5 Potential environmental effects. Appropriate questions for specific applications to be added by users of the framework.

Specific criteria for evaluating control of the microorganism after it is introduced must include the potentials for persistence of the introduced microorganism, genetic exchange between the introduced and indigenous microorganisms, and spread of the introduced microorganism to nontarget environments (Fig. 11-3). A series of questions to be addressed in evaluating the potential for unwanted persistence of an introduced microorganism is illustrated in Fig. 11-4.

Criteria for evaluating effects must depend, at least in part, on the intended function of the introduced microorganism in its target environment (Fig. 11-5). Thus, a proposed field test of a bacterium to be used for biodegradation of a toxic pollutant should be preceded by definitive laboratory experiments and should be designed to determine whether toxic by-products of the degradation may be created and persist.

As the agencies grant permission to introduce genetically modified microorganisms in field tests, they will receive advice from panels of experts who can utilize the decision framework described here. With experience, familiarity will increase, and we anticipate this will be accompanied by adjustments in the rigor of oversight.

Appendix
Historical Overview of Nucleic Acid Biotechnology: 1973 to 1989

The origins of current initiatives by federal agencies to regulate planned introductions of genetically manipulated organisms, particularly those derived from recombinant DNA techniques, lie in the concerns of scientists who recognized in the early 1970s that the ability to specifically alter the genetic code has far-reaching implications. Since then, the waxing and waning of interest by Congress, federal agencies, states, and local municipalities in regulation of modern biotechnological methods and products has paralleled changing perceptions of risk (Korwek and de la Cruz, 1985).

Oversight mechanisms tailored to the methods and products of biotechnology began to emerge in 1974, when the National Academy of Sciences (NAS), responding to a letter from the attendees at the 1973 Gordon Conference on Nucleic Acids, convened a committee to evaluate the safety of research on recombinant DNA. The committee published its recommendations in *Nature* (Berg et al., 1974a) and *Science* (Berg et al., 1974b) calling for a voluntary moratorium on recombinant DNA experiments while questions of public safety were further evaluated. The letter also invited the National Institutes of Health (NIH) to establish a committee to oversee an evaluation of potential biological and ecological hazards and to devise guidelines for working with recombinant DNA.

The debate over safety concerns was extended to include broader social issues at the February 1975 International Conference on Recombinant DNA Molecules (often called the Asilomar Conference), which was convened at the Asilomar Conference Center, California, by the Assembly of Life Sciences of the National Research Council. Participants at the Asilomar Conference also debated the ethical issues raised by recombinant DNA research as well as the legal liabilities of the investigators and institutions in the event of injury arising from such research (Berg et al., 1975a; Berg et al., 1975b). Some participants argued that recombinant DNA research should proceed unrestricted by guidelines or special regulations, while others maintained that the potential dangers demanded restrictions or self-imposed guidelines. Ultimately, a statement of principles outlining a proposed set of standards for recombinant DNA research was drafted, and researchers agreed to control their own research stringently until the safety of the new recombinant DNA technology could be ensured.

FORMATION OF NIH GUIDELINES

A second stage in the development of an oversight mechanism began when a committee of scientists appointed by NIH, known as the Recombinant DNA Advisory Committee (RAC), converted the statement of principles developed at the Asilomar Conference into research- and containment-oriented guidelines. The first guidelines for research involving recombinant DNA molecules were published in 1976 for use in overseeing NIH-funded research activities (NIH, 1976). Their initial focus was on containment designed to ensure the safety of laboratory work and to prevent the accidental escape of recombinant DNA microorganisms. Risk categories for experiments were assigned; different types of experimental work were to be conducted at different levels of physical and biological containment; and other experiments, including environmental introductions, were prohibited.

As experience with contained applications of recombinant DNA accumulated, many of the risks feared to be associated with laboratory recombinant DNA research were found to be greatly overestimated or simply nonexistent (Levin, 1984). As a result, in 1978 the standards of containment required for a range of recombinant DNA experiments conducted in the laboratory were relaxed (NIH, 1978). Subsequent revisions have included decentralization of responsibility for the administration of recombinant DNA experiments, simplification of the administrative procedures for working with recombinant

DNA, and, to prevent duplicative review, exemption from RAC and NIH review for certain experiments submitted for review to a federal regulatory agency. Increased responsibility for oversight of recombinant DNA research also has been placed in the hands of local institutional biosafety committees. The scope of the guidelines also has been expanded from a focus on research to a concern with large-scale operations, from in vitro work to possible applications of gene therapy to humans, and from laboratory containment to environmental introductions (Vandenbergh, 1986; Korwek, 1988).

The guidelines eventually became binding on all institutions receiving any federal funding, in addition to those receiving NIH grants, and their influence has spread beyond federally funded research activities and beyond application of recombinant DNA techniques. In the period since their adoption, state and local governments, academic institutions, the industrial community, and foreign countries have voluntarily applied the guidelines or modified versions of them. In addition, the RAC, which has been expanded to include persons in a variety of disciplines, has served as a model for the formation of biotechnology advisory groups for federal regulatory agencies. The Environmental Protection Agency's Biotechnology Science Advisory Committee (BSAC) and the U.S. Department of Agriculture's (USDA's) Biotechnology Research Advisory Committee (ABRAC) are examples of groups modeled after the RAC. These groups provide advice on scientific and policy issues involving agency oversight of a broad range of technologies, in addition to recombinant DNA.

ENVIRONMENTAL RELEASES

The modification of the guidelines to address the planned introduction into the environment of certain genetically manipulated organisms triggered another stage in the development of an oversight system and a new debate about hazards. Progress in research during the mid-1970s permitted the development of genetically manipulated microorganisms designed to survive and function outside the laboratory. As a result, the guidelines were amended in 1978 to continue the general prohibition on planned introductions, but to permit the NIH director, on the advice of the RAC, to grant exceptions (NIH, 1978).

Three requests between 1980 and 1983 to field-test plants and microorganisms containing recombinant DNA forced the RAC to move from the ad hoc approach outlined in the 1978 guideline amendments to the creation of generally applicable release guidelines, but they also forced the development of federal regulatory initiatives. The first of these requests was made by Stanford University researchers in March 1980 to test maize (*Zea mays*) transformed by DNA cloned from *Escherichia coli* and the yeast *Saccharomyces cerevisiae* in an attempt to modify zein, a grain-storage protein. Cornell University next requested a field test for tomatoes and tobacco seedlings developed from pollen containing DNA from a hybrid plasmid vector carrying antibiotic resistance markers. Although both tests were approved by NIH and USDA (NIH, 1981; NIH, 1983), neither was carried out.

The third request, in September 1982, came from researchers at the University of California at Berkeley and proved to be the most controversial. The RAC reviewed a proposal to field-test the plant bacteria *Pseudomonas syringae* subsp. *syringae* and *Erwinia herbicola* with deletions of genetic information for the ice nucleation factor. The RAC requested that a revised version of the proposal to test these "ice-minus" bacteria be prepared. It reviewed the revised proposal in October 1982 and approved it seven months after submission of the initial request (NIH, 1983).

NIH approval of the ice-minus experiment then provoked a court challenge under the National Environmental Policy Act (NEPA) (U.S. Congress, 1982). NEPA establishes procedures obligating many federal agencies to take environmental values into account for all major activities. It requires most federal agencies to conduct an environmental assessment and perhaps to prepare an environmental impact statement for each major action that may significantly affect the environment. A federal district court enjoined the field test of *Pseudomonas* and *Erwinia* isolates on the ground that the RAC review did not adequately consider the environmental impacts of the release of these particular ice-minus microorganisms (*Foundation on Economic Trends v. Heckler*, 1984). The court enjoined NIH from approving future environmental release proposals on the ground that the RAC approval process required a programmatic environmental impact statement under NEPA. A federal appeals court subsequently reversed the district court's ruling requiring such an impact statement, but upheld the injunction against the ice-minus experiment pending NEPA review (*Foundation on Economic Trends v. Heckler*, 1985). This case established a precedent for further NEPA challenges to other applications of modern methods of nucleic acid biotechnology (*Foundation on Economic Trends v. Lyng*, 1986).

CONGRESSIONAL INITIATIVES

The development of genetically modified microorganisms designed to function outside the laboratory has also prompted several congressional hearings into the environmental hazards of planned introductions and the adequacy of regulatory oversight mechanisms. No specific legislation has been enacted. In June 1983, Congressmen Douglas Walgren (chairman of the Subcommittee on Science, Research and Technology) and Albert Gore (chairman of the Subcommittee on Investigations and Oversight) conducted a hearing on the environmental implications posed by commercial applications of recombinant DNA technology (U.S. Congress, 1983). This hearing followed the United States Supreme Court's decision in *Diamond v. Chakrabarty* (U.S. Congress, 1980; Wade, 1980), which upheld the patentability of life forms and provided a stimulus to the commercial development of genetically manipulated microorganisms for both laboratory and nonlaboratory use.

The report of the hearing concluded that predicting the environmental effects from the introduction of genetically manipulated organisms is difficult, but that any highly negative consequence had a low probability of occurring (U.S. Senate, 1984). The report also questioned the ability of federal agencies to regulate planned introductions in light of the unquantifiable nature of the risks, and it concluded that more information on the environmental fate of these introduced organisms was needed to ascertain whether such releases posed a risk to the ecosystem.

Similarly, in September 1984, the Subcommittee on Toxic Substances and Environmental Oversight of the Senate Committee on Environmental and Public Works held a hearing on "the potential environmental consequences of genetic engineering" (U.S. Senate, 1984). Representatives from the Environmental Protection Agency (EPA), NIH, and USDA testified that existing statutes, regulations, and guidelines would benefit from clarification, but were adequate to address release issues without congressional intervention. Before the Senate hearing, an interagency working group was formed under the White House Cabinet Council on Natural Resources and the Environment to review biotechnology regulation and to begin the process of coordinating the biotechnology activities of the federal agencies.

DEVELOPMENT OF REGULATORY OVERSIGHT

In December 1984, the working group proposed a regulatory

strategy including a matrix of laws applicable to biotechnology; it also included individual policy statements from USDA, EPA, and the Food and Drug Administration (FDA) outlining their regulatory roles (OSTP, 1984). The working group also proposed the formation of a scientific biotechnology science board to coordinate regulatory activities of the different agencies and to provide advice on scientific issues related to biotechnology.

In response to criticism that the Biotechnology Science Board would further complicate an already complex regulatory system, the board was replaced in October 1985 when the Biotechnology Science Coordinating Committee (BSCC) was created to develop a common scientific approach within the coordinated federal regulatory framework for biotechnology (OSTP, 1985). In addition, the responsibilities for biotechnology coordination within the Reagan administration were shifted to the Domestic Policy Council Working Group on Biotechnology within the Office of Science and Technology Policy (OSTP, 1985).

In June 1986, OSTP published the Coordinated Framework for Regulation of Biotechnology, which identifies the agencies responsible for approving biotechnology products and their respective jurisdictions for regulating planned introductions (OSTP, 1986). Overall, the coordinated framework reiterates the earlier view that the current laws are adequate to oversee current biotechnology developments. Since the possibility of regulatory overlap exists, particularly among EPA, FDA, and USDA, the document identifies which regulatory bodies have been designated as lead agencies for particular biotechnology products or their uses.

Although the current oversight framework is still evolving, the regulatory agencies continue to rely on existing laws for oversight of biotechnology activities. Under existing statues and the 1986 Coordinated Framework, products of biotechnology and research and commercial applications may be regulated differently and by different agencies. Variables that may trigger regulatory oversight include the extent of the genetic manipulation and the intended use of the product, for example, whether a product is to be used as a pesticide, food, or drug. In other respects oversight depends on whether a plant or animal pathogen may be involved. Some laws prevent duplication of federal and state review while others do not, thus leading to the possibility of oversight of biotechnology by more than one federal or state agency. This diversity in the bases for regulation and in the oversight mechanisms is derived from the variety of federal and state laws that have been enacted to protect human health and the environment.

The current oversight framework is based on the authority in various laws to require permits or other types of agency review *before* introduction. It may be briefly characterized by (1) attempts by federal and state agencies to coordinate their regulatory activities (with various degrees of success), to prevent overlapping regulation; (2) reliance on outside committees, such as BSAC and ABRAC, for review of scientific and other issues; (3) ongoing efforts to modify and refine existing regulatory mechanisms; (4) a case-by-case approach, especially in reviewing proposed field tests; and (5) a shifting emphasis from scrutiny of only the processes utilized (for example, recombinant DNA techniques) to scrutiny of the characteristics of the derived products.

Recurrent difficulties in the oversight of planned introductions have involved a variety of considerations, including whether adequate scientific bases exist for the federal agencies to differentiate releases of greater and lesser concern, whether data requirements are appropriate, and whether emerging regulatory approaches (which tend to be product- rather than process-based) will extend the reach of oversight to areas not traditionally subject to federal review.

With the shift of focus from process to product, a new problem has arisen: Regulatory oversight might be triggered not only for new nucleic acid technologies, but also for those that have not been manipulated at all or that have been developed through classical techniques, such as mutagenesis. As a result, the product-based approach engenders the possibility that planned introductions of products of older technologies may also become subject to special oversight, in many cases for the first time, perhaps even despite a long history of safe use.

With respect to risk-assessment issues, a fundamental concern is whether the limited current understanding of microbial ecology (McGarity and Bayer, 1983; Strauss et al., 1986) enables the environmental fate of released organisms to be predicted. The oversight of planned introductions by NIH and the federal regulatory agencies can generally be described as science-based, and the more than 50 releases that have been permitted thus far have been allowed because of their perceived "low-risk" status, in light of the characteristics of the genetically manipulated organism and the small scale of the field-test environment into which it is introduced.

Although the data requirements of the federal regulatory agencies are not identical, they share several common features. Where DNA has been moved from one organism to another, each agency usually requires information about the parent or source organism and its characteristics, the identity and function of the genetic material transferred, and the mechanism by which the DNA was transferred. The agencies also require information on the organism that is the subject of the genetic work, including data on the characteristics expressed before and after manipulation, such as the likelihood of competitive success in the environment and of subsequent genetic transfer to other organisms.

The agencies usually also require data on the characteristics of the planned introduction, such as the environment into which the genetically manipulated organism will be released, the size of the release area, and the number of organisms to be introduced. Release requirements often include containment principles that will limit the proliferation of the introduced organism, such as the limiting characteristics of the organism itself and other biological and physical mechanisms that help prevent dissemination beyond the test site. Finally, the agencies have attempted to retain the flexibility to require additional data where needed.

EVALUATION OF OVERSIGHT CAPABILITIES

Several recent reports have discussed the regulatory regime, especially risk-assessment capabilities and the adequacy of oversight mechanisms. In September 1987, an NAS white paper (NAS, 1987) concluded that there is no evidence that the introduction into the environment of organisms modified by recombinant DNA present unique hazards, but rather that the risks are the same that as incurred in the introduction into the environment of unmodified organisms. Consistent with the oversight approaches sometimes utilized by the regulatory agencies and others, the white paper concluded that decision-making on the environmental use of genetically manipulated organisms should be based on the organisms' relevant properties and not on the process by which the organisms were produced. It also recommended that the scientific community provide guidance to assist investigators and regulators in evaluating the planned introduction of modified organisms from an ecological perspective (Tiedje et al., 1989).

A May 1988 report by the Office of Technology Assessment

(OTA) illustrated a range of options for congressional action in major areas of public policy, including the criteria for review of planned introductions for potential risk, the administrative mechanisms for applying such review criteria, and the research base supporting planned introductions (OTA, 1988). The OTA report concluded that although reasons exist to continue to be cautious about environmental introductions, there is no cause for alarm. The report also noted that some questions can be answered only with practical experience, that is, with realistic small-scale field tests, which are not likely to result in environmental problems. It also called for the establishment of broad categories that can be used to sort proposed introductions for low, medium, or high levels of review.

At the behest of the Subcommittee on Oversight and Investigations of the House Committee on Energy and Commerce, the General Accounting Office (GAO) issued a report in June 1988 reviewing the federal risk management of genetically engineered organisms intended for agricultural and health use in the environment (GAO, 1988). The report evaluated the scope of regulatory policies applicable to small-scale releases, reviewed the administrative procedures for implementing policies, and identified technical methods available to control and monitor risks posed by field testing. It states that the probability of ecological disruption from introductions is low, but the magnitude of the impact may be extremely severe. The report also notes that USDA, EPA, and FDA have made efforts to coordinate their policies and review procedures, but they have limited experience with genetically manipulated organisms used in the environment and are uncertain about the effects of the organisms. Recommendations include the elimination of certain classes of introductions that are currently exempt from federal agency review because of the incompleteness of the scientific underpinning needed to justify these exemptions.

Literature Cited

Anderson, E. 1949. Introgressive Hybridization. New York: Wiley. 109 pp.

Anderson, E. 1952. Plants, Man and Life. Boston: Little, Brown. 245 pp.

Association of Official Seed Certification Agencies. 1971. AOSCA Certification Handbook. Publication 23.

Attwood, G. T., R. A. Lockington, G. P. Xue, and J. D. Brooker. 1988. Use of unique gene sequences as a probe to enumerate a strain of *Bacteroides rumincola* introduced into the rumen. Appl. Environ. Microbiol. 54:534-539.

Austin, S., M. A. Baer, and J. P. Helgeson. 1985. Transfer of resistance to potato leaf roll virus from *Solanum brevidens* into *Solanum tuberosum* by somatic fusion. Plant Sci. 39:75-82.

Ayres, J. C., J. O. Mundt, and W. E. Sandine. 1980. Microbiology of Foods. San Francisco: W.H. Freeman. 708 pp.

Bahme, J. B., and M. N. Schroth. 1987. Spatial-temporal colonization patterns of a rhizobacterium on underground organs of potato. Phytopathology. 77:1093-1100.

Baker, H. G. 1972. Migration of weeds. Pp. 327-347 in Taxonomy, Phytogeography and Evolution, D. H. Valentine, ed. London: Academic Press.

Baker, H. G. 1974. The evolution of weeds. Annu. Rev. Ecol. Syst. 5:1-24.

Baker, H. G. 1986. Patterns of plant invasion in North America. Pp. 44-57 in Ecology of Biological Invasions of North America and Hawaii, H. A. Mooney and J. A. Drake, eds. Ecological Studies 58. New York: Springer-Verlag.

Banwart, G. J. 1979. Basic Food Microbiology. Westport, Conn.: AVI Publishing. 781 pp.

Barrett, J. 1985. Gene-for-gene hypothesis: Parable or paradigm. Pp. 215-225 in Ecology and Genetics of Host-Parasite Interaction, D. Rollinson and R. M. Anderson, eds. London: Academic Press.

Barrett, S. C. H. 1983. Crop mimicry in weeds. Econ. Bot. 37:255-282.

Barrett, S. C. H., and B. J. Richardson. 1986. Genetic attributes of invading species. Pp. 21-23 in Ecology of Biological Invasions, R. H. Groves and J. J. Burdon, eds. Cambridge, England: Cambridge University Press.

Barry, G. F. 1986. Permanent insertion of foreign genes into the chromosomes of soil bacteria. Bio/Technology 4:446-449.

Bazzaz, F. A. 1986. Life history of colonizing plants: Some demographic, genetic, and physiological features. Pp. 96-110 in Ecology of Biological Invasions of North America and Hawaii, H. A. Mooney and J. A. Drake, eds. Ecological Studies: Vol 58. New York: Springer-Verlag.

Beachy, R. N., Z. L. Chen, R. B. Horsch, S. G. Rogers, N. J. Hoffmann, and R. T. Fraley. 1985. Accumulation and assembly of soybean-conglycinin in seeds of transformed petunia plants. EMBO J. 4:3047-3053.

Berg, P., D. Baltimore, and H. W. Boyer. 1974a. NAS ban on plasmid engineering. Nature 250:175.

Berg, P., D. Baltimore, and H. W. Boyer. 1974b. Potential biohazards of recombinant DNA molecules. Science 185:303.

Berg, P., D. Baltimore, and S. Brenner. 1975a. Asilomar conference on recombinant DNA molecules. Science 188:991-994.

Berg, P., D. Baltimore, and S. Brenner. 1975b. Summary statement of the Asilomar conference on recombinant DNA molecules. Proc. Natl. Acad. Sci. U.S.A. 72:1981-1984.

Bevan, M. 1984. Agrobacterium vectors for plant transformation. Nucleic Acids Res. 12:8711-8718.

Bishop, D. H. L. 1986. UK release of genetically marked virus. Nature 323:496.

Bishop, D. H. L. 1988. The release into the environment of genetically engineered viruses, vaccines and viral pesticides. Pp. 12-15 in Planned Release of Genetically Engineered Organisms (Trends in Biotechnology/Trends in Ecology and Evolution Special Issue), J. Hodgson and A. M. Sugden, eds. Cambridge, England: Elsevier.

Bishop, D. H. L., P. F. Entwistle, I. R. Cameron, C. J. Allen, and R. D. Possee. 1988. Field trials of genetically engineered baculovirus insecticides. Pp. 143-179 in The Release of Genetically-Engineered Micro-Organisms, M. Sussman, G. H. Collins, F. A. Skinner, and D. E. Stewart-Tall, eds. London, England: Academic Press.

Bohlool, B. B., and E. L. Schmidt. 1980. The immunofluoresence approach in microbial ecology. Adv. Microbial Ecol. 4:203-241.

Bopp, L. H. 1986. Degradation of highly chlorinated PCB's by *Pseudomonas* strain LB400. J. Ind. Microbiol. 1:23-29.

Borenstein, S., and E. Ephrati-Elizur. 1969. Spontaneous release of DNA in sequential genetic order by *Bacillus subtilus*. J. Mol. Biol. 45:137-152.

Bouma, J. E., and R. E. Lenski. 1988. Evolution of a bacteria/plasmid association. Nature 335:351-352.

Bouwer, E. J., B. E. Rittman, and P. L. McCarty. 1981. Anaerobic degradation of halogenated 1- and 2-carbon organic compounds. Environ. Sci. Technol. 15:596-599.

Boyce Thompson Institute for Plant Research. 1987. Regulatory Considerations: Genetically Engineered Plants. San Francisco: Center for Science Information. 106 pp.

Boynton, J. E., N. W. Gillham, E. H. Harris, J. P. Hosler, A. M. Johnson, A. R. Jones, B. L. Randolph-Anderson, D. Robertson, T. M. Klein, K. B. Shark, and J. C. Sanford. 1988. Chloroplast transformation in Chlamydomonas with high-velocity microprojectiles. Science 240:1534-1538.

Braun, V., and K. Hantke. 1977. Bacterial receptors for phages and colicins as constituents of specific transport systems. Pp. 99-137 in Microbial Interactions, J. L. Reissig, ed. London: Chapman & Hall.

Brill, W. J. 1985. Safety concerns and genetic engineering in agriculture. Science 227:381-384.

Brisson, N., J. Paszkowski, J. R. Penswick, B. Gronenborn, I. Potrykus, and T. Hohn. 1984. Expression of a bacterial gene in plants by using a viral vector. Nature 310:511-514.

Brisson-Noel, A., M. Arthur, and P. Courvalin. 1988. Evidence for natural gene transfer from gram-positive cocci to *Escherichia coli*. J. Bacteriol. 170:1739-1745.

Brown, M. P., R. E. Wagner, and D. L. Bernard. 1988. PCB dechlorination in Hudson River sediment. Science 240:1675-1676.

Brubaker, R. R. 1985. Mechanisms of bacterial virulence. Annu. Rev. Microbiol. 39:21-50.

Burdon, J. J., R. H. Groves, and J. M. Cullen. 1981. The impact of biological control on the distribution and abundance of *Chondrilla juncea* in southeastern Australia. J. Appl. Ecol. 18:957-966.

Cairns, J., Jr., M. Alexander, K. W. Cummins, W. T. Edmondson, R. Goldman, J. Harte, A. R. Insensee, R. Levin, J. F. McCormick, T. J. Peterle, and J. H. Zar. 1981. Testing for Effects of Chemicals on Ecosystems. Washington, D.C.: National Academy Press.

Chase, S. S. 1969. Monoploids and monoploid derivatives of maize (*Zea mays* L.). Bot. Rev. 35:117-167.

Choo, T. M., E. Reinbergs, and K. J. Kasha. 1985. Use of haploids in breeding barley. Plant Breeding Rev. 3:219-252.

Close, T., and R. L. Rodriguez. 1982. Construction and characterization of the chloramphenicol resistance gene cartridge: A new approach to the transcriptional mapping of extrachromosomal elements. Gene 20:305-316.

Collmer, A. S., and N. T. Keen. 1986. The role of pectic enzymes in plant pathogenesis. Annu. Rev. Phytopathol. 24:383-409.

Colwell, R. K., E. A. Norse, D. Pimentel, F. E. Sharples, and D. Simberloff. 1985. Genetic engineering in agriculture. Science 229:111-112.

Colwell, R. R., P. R. Brayton, D. J. Grimes, D. B. Roszak, S. A. Huq, and L. M. Palmer. 1985. Viable but non-culturable *Vibrio cholerae* and related pathogens in the environment: Implications for the release of genetically engineered microorganisms. Bio/Technology 3:817-820.

Colwell, R. R., C. Somerville, I. Knight, and W. Straube. 1988. Detection and monitoring of genetically engineered micro-organisms. Pp. 47-60 in The Release of Genetically Engineered Micro-Organisms, M. Sussman, G. H. Collins, F. A. Skinner, and D. E. Stewart-Tall, eds. London: Academic Press.

Costerton, J. W., R. T. Irvin, and K. J. Chang. 1981. The bacterial glycocalyx in nature and disease. Annu. Rev. Microbiol. 35:299-324.

Cripe, C. R., and P. H. Pritchard. In preparation. Microcosms as Test Systems for Fate and Transport of Microorganisms.

Crossway, A., H. Hauptli, C. M. Houck, J. M. Irvine, J. V. Oakes, and L. A. Perani. 1986. Micromanipulation techniques in plant biotechnology. BioTechniques 4:320-334.

Crow, J. F., and M. Kimura. 1970. An introduction to population genetics theory. New York: Harper & Row. [Reprinted Minneapolis: Burgess].

Crow, M. E., and F. B. Taub. 1979. Designing a microcosm bioassay to detect ecosystem level effects. Intern. J. Environ. Stud. 13:141-147.

Curtiss, R., III. 1988. Engineering organisms for safety: What is necessary? Pp. 7-20 in The Release of Genetically Engineered Micro-Organisms, M. Sussman, G. H. Collins, F. A. Skinner, and D. E. Stewart-Tall, eds. London: Academic Press.

Davis, B. D. 1987. Bacterial domestication: Underlying assumptions. Science 235:1329-1335.

Davison, J. 1988. Plant beneficial bacteria. Bio/Technology 6:282-286.

de Wet, J. M. J. 1966. The origin of weediness in plants. Proc. Oklahoma Acad. Sci. 47:14-17.

Decker, D., and H. D. Wilson. 1987. Allozyme variation in the *Cucurbita pepo* complex: *C. pepo* var. *ovifera* vs. *C. texana*. Syst. Bot. 12:263-273.

DeMaagd, R. A., R. DeRijk, I. H. M. Mulders, and B. J. J. Logtenberg. 1989. Immunological characterization of *Rhizobium leguminosarum* outer membrane antigens by use of polyclonal and monoclonal antibodies. J. Bacteriol. 171: 1136-1142.

Deroles, S. C., and R. C. Gardner. 1988. Expression and inheritance of kanamycin resistance in a large number of transgenic petunias generated by *Agrobacterium*-mediated transformation. Plant Mol. Biol. 11:355-364.

Dessaux, Y., J. Tempé, and S. K. Farrand. 1987. Genetic analysis of mannitol opine catabolism in octopine-type *Agrobacterium tumefaciens* strain 15955. Mol. Gen. Genet. 208:301-308.

Dewey, R. E., D. H. Timothy, and C. S. Levings III. 1987. A mitochondrial protein associated with cytoplasmic male sterility in the T cytoplasm of maize. Proc. Natl. Acad. Sci. U.S.A. 84:5374-5378.

Dewey, R. E., J. N. Siedew, D. H. Timothy, and C. S. Levings III. 1988. A 13-kilodalton maize mitochondrial protein in *E. coli* confers sensitivity to *Bipolaris maydis* toxin. Science 239:293-295.

Diaz, R. J., M. Luckenbach, S. Thornton, M. H. Roberts, Jr., R. J. Livingston, C. C. Koenig, G. L. Ray, and L. E. Wolfe. 1987. Field validation of multi-species laboratory test systems for estuarine benthic communities: Project Summary EPA/600/S3-87/016. Gulf Breeze, Fla.: U.S. Environmental Protection Agency, Environmental Research Laboratory.

Djordjevic, M. A., D. W. Gabriel, and B. G. Rolfe. 1987. Rhizobium: The refined parasite of legumes. Annu. Rev. Phytopathol. 25:145-168.

Djordjevic, M. A., J. W. Redmond, M. Batley, and B. G. Rolfe. 1987. Clover secretes specific phenolic compounds which either stimulate or repress *nod* gene expression in *Rhizobium trifolii*. EMBO J. 6:1173-1179.

Doebley, J. F. 1984. Maize introgression into teosinte—A reappraisal. Ann. Mo. Bot. Gard. 71:1100-1113.

Dowling, D. N., and W. J. Broughton. 1986. Competition for nodulation of legumes. Annu. Rev. Microbiol. 40:131-157.

Downey, R. K., and J. F. W. Rakow. 1987. Rapeseed and mustard. Pp. 437-486 in Principles in Cultivar Improvement, W. R. Fehr, ed. New York: MacMillian.

Drahos, D. J., B. C. Hemming, and S. McPherson. 1986. Tracking recombinant organisms in the environment: β-galactosidase as a selectable non-antibiotic marker for fluorescent pseudomonads. Bio/Technology 4:439-444.

Drahos, D. J., G. F. Barry, B. C. Hemming, E. J. Brandt, H. D. Skipper, E. L. Kline, D. A. Kluepfel, T. A. Hughes, and D. T. Gooden. 1988. Pre-release testing procedures: U.S. field test of a lacZY-engineered soil bacterium. Pp. 181-191 in The Release of Genetically Engineered Micro-Organisms, M. Sussman, G. H. Collins, F. A. Skinner, and D. E. Stewart-Tall, eds. London: Academic Press.

Duncan, K. E., C. A. Istock, and N. Fergason. In preparation. Genetic exchange between *Bacillus subtilus* and *Bacillus licheniformis*: Variable hybrid stability and the nature of bacterial species.

Duval-Iflah, Y., P. Raibaud, and M. Rousseau. 1981. Antagonisms among isogenic strains of *Escherichia coli* in the digestive tracts of gnotobiotic mice. Infect. Immunol. 34:957-969.

Dykhuizen, D., and D. L. Hartl. 1980. Selective neutrality of 6PGD allozymes in *E. coli* and the effects of genetic background. Genetics 96:801-817.

Edlin, G., R. C. Tait, and. R. L. Rodriguez. 1984. A bacteriophage lambda cohesive ends (cos) DNA fragment enhances the fitness of plasmid-containing bacteria growing in energy-limited chemostats. Bio/Technology 2:251-254.

Ehlenfeldt, M. K., and J. P. Helgenson. 1987. Fertility of somatic hybrids from protoplast fusion of *Solanum brevidens* and *S. tuberosum*. Theor. Appl. Genet. 73:395-402.

Ehrlich, H. L., ed. 1981. Geomicrobiology. New York: Marcel Dekker. 392 pp.

Ellingboe, A. H. 1982. Genetical aspects of active defence. Pp. 192-197 in Active Defense Mechanisms in Plants, R. K. S. Wood, ed. London: Plenum.

Ellis, J. G., A. Kerr, M. Van Montagu, and J. Schell. 1979. *Agrobacterium* genetic studies on agrocin 84 production and the biological control of crown gall. Physiol. Plant Pathol. 15:311-319.

Ellis, J. G., P. J. Murphy, and A. Kerr. 1982. Isolation and properties of transfer regulatory mutants of the nopaline Ti plasmid pTiC58. Mol. Gen. Genet. 186:275-281.

Ellstrand, N. C. 1988. Pollen as a vehicle for the escape of engineered genes? Trends Biotechnol. 6:530-532.

Entwistle, P. F., and H. F. Evans. 1985. Viral control. Pp. 347-412 in Comprehensive Insect Physiology, Biochemistry and Pharmacology, vol. 12, L. I. Gilbert and G. A. Kerkut, eds. Oxford: Pergamon.

Falcone, G., M. Campa, H. Smith, and G. M. Scott, eds. 1984. Bacterial and Viral Inhibition and Modulation of Hosts Defenses. London: Academic Press. 249 pp.

Farrand, S. K., J. E. Slota, J. S. Shim, and A. Kerr. 1985. Tn5 insertions in the agrocin 84 plasmid: The conjugal nature of pAgK84 and the locations of determinants for transfer and agrocin 84 production. Plasmid 13:106-117.

Faust, M. A. 1982. Relationship between land-use practices and fecal bacteria in soils. J. Environ. Qual. 11:141-146.

Federation of British Plant Pathologists. 1973. A guide to the use of terms in plant pathology. Phytopathology 17:1-55.

Feldman, K. A., M. D. Marks, M. L. Christianson, and R. S. Quatrano. 1989. A dwarf mutant of *Arabidopsis* generated by T-DNA insertion mutagenesis. Science 243:1351-1354.

Firmin, J. J., K. E. Wilson, L. Rossen, and A. W. B. Johnston. 1986. Flavonoid activation of nodulation genes in *Rhizobium* reversed by other compounds present in plants. Nature 324:90-92.

Fischhoff, D. A., K. S. Bowdish, F. J. Perlak, P. G. Marrone, S. M. McCormick, J. G. Niedermeyer, D. A. Dean, K. Kusano-Krettzmer, E. J. Mayer, D. E. Rochester, S. G. Rogers, and R. T. Fraley. 1987. Insect-tolerant transgenic tomato plants. Bio/Technology. 5:807-813.

Fliermans, C. B., T. J. Phelps, D. Ringelberg, A. T. Mikell, and D. C. White. 1988. Mineralization of trichloroethylene by heterotrophic enrichment cultures. Appl. Environ. Microbiol. 54:1709-1714.

Flor, H. H. 1956. The complementary genic systems in flax and flax rust. Adv. Genet. 8:29-54.

Flor, H. H. 1958. Mutations to wider virulence in *Melampsora lini*. Phytopathology 48:297-301.

Focht, D. 1988. Performance of biodegradative microorganisms in soil: Xenobiotic chemicals as unexploited metabolic niches. Pp. 15-29 in Environmental Biotechnology: Reducing Risks from Environmental Chemicals through Biotechnology, G. S. Omenn, ed. New York: Plenum.

Focht, D. D., and W. Brunner. 1985. Kinetics of biphenyl and polychlorinated biphenyl metabolism in soil. Appl. Environ. Microbiol. 50:1058-1063.

Forde, B. G., and C. J. Leaver. 1989. Nuclear and cytoplasmic genes controlling synthesis of variant mitochondrial polypeptides in male sterile maize *Zea mays*. Proc. Natl. Acad. Sci. U.S.A. 77:418-422.

Forseth, I. N., and A. H. Teramura. 1987. Field photosynthesis, microclimate and water relations of an exotic temperate liana, *Puerania lobata*, kudzu. Oecologia 71:262-267.

Foundation of Economic Trends v. Heckler, 587 F. Supp. 753 (D.D.C. 1984), aff'd in part. vacated in part. 756 F.2d 143 (D.C. Cir. 1985).

Foundation on Economic Trends v. Heckler, 756 F.2d 143 (D.C. Cir. 1985).

Foundation on Economic Trends v. Lyng, Civ. No. 86-1130 (D.D.C., filed April 26, 1986).

Fraley, R. T., S. G. Rogers, and R. B. Horsch. 1986. Genetic transformation in higher plants. Crit. Rev. Plant Sci. 4:1-46.

Frantz, B., and A. M. Chakrabarty. 1986. Degradative plasmids in *Pseudomonas*. Pp. 295-323 in The Bacteria, vol. 10, J. R. Sokatch and L. N. Ornston, eds. New York: Academic Press.

Fredrickson, J. K., D. F. Bezdicek, F. J. Brockman, and S. W. Li. 1988. Enumeration of Tn*5* mutant bacteria in soil by using a most-probable-number DNA hybridization procedure and antibiotic resistance. Appl. Env. Microbiol. 54: 446-453.

Freifelder, D. 1987. Microbial Genetics. Boston: Jones & Bartlett. 601 pp.

French, R., M. Janda, and P. Ahlquist. 1986. Bacterial gene inserted in an engineered RNA virus: Efficient expression in monocotyledonous plant cells. Science 231:1294-1297.

Frick, T. D., R. L. Crawford, M. Martinson, T. Chresand, and G. Bateson. 1988. Microbiological cleanup of groundwater contaminated by pentachlorophenol. Pp. 173-191 in Environmental Biotechnology: Reducing Risks from Environmental Chemicals through Biotechnology, G. S. Omenn, ed. New York: Plenum.

Fromm, M. E., L. P. Taylor, and V. Walbot. 1986. Stable transformation of maize after gene transfer by electroporation. Nature 319:791-793.

Fromm, M. E., L. P. Taylor, and V. Walbot. 1987. Electroporation of DNA and RNA into plant protoplasts. Methods Enzymol. 153:351-366.

Fry, J., A. Barnason, and R. B. Horsch. 1987. Transformation of *Brassica napus* with *Agrobacterium tumefaciens* based vectors. Plant Cell Rep. 6:321-325.

Fulbright, D. 1989. In press. Molecular basis for hypovirulence and its ecological relationships. In New Directions in Biological Control, R. Baker, ed. New York: Alan R. Liss.

Furner, I. J., G. A. Huffman, R. A. Amasino, D. J. Garfinkel, M. P. Gordon, and E. W. Nester. 1986. An *Agrobacterium* transformation in the evolution of the genus *Nicotiana*. Nature 319:422-427.

Gabriel, D. W., A. Burges, and G. R. Lazo. 1986. Gene-for-gene interactions of five cloned avirulence genes from *Xanthomonas campestris* pv. *malvacearum* with specific resistance genes in cotton. Proc. Natl. Acad. Sci. U.S.A. 83:6415-6419.

Gade, D. W. 1976. Naturalization of plant aliens: the volunteer orange in Paraguay. J. Biogeogr. 3:269-279.

Gasser, C. S., and R. T. Fraley. 1989. Genetically engineered plants for crop improvement. Science 244:1293-1299.

GAO (U.S. General Accounting Office). 1988 (June). Biotechnology: Managing Risks of Field Testing Genetically Engineered Organisms. Washington, D.C.: Government Printing Office (GAO/RCED-88-27). 108 pp.

Ghosal, D., I. S. You, D. K. Chatterjee, and A. M. Chakrabarty. 1985. Microbial degradation of halogenated compounds. Science 228:135-142.

Gibson, R. W., M. G. K. Jones, and N. Fish. 1988. Resistance to potato leaf roll virus and potato virus Y in somatic hybrids between dihaploid *Solanum tuberosum* and *S. brevidens*. Theor. Appl. Genet. 76:113-117.

Giesy, J. P., Jr., ed. 1980. Microcosms in Ecological Research. Springfield, Va.: National Technical Information Service; DOE Symposium 52.

Gillett, J. W., and J. M. Witt, eds. 1979. Terrestrial Microcosms. Washington, D.C.: National Science Foundation. NSF-RA-79-0034. 34 pp.

Gillett, J. W., J. D. Gile, and L. K. Russell. 1983. Predator-prey (vole-cricket) interactions: Effects of wood preservatives. Environ. Toxicol. Chem. 2:83-93.

Gillett, J. W., A. M. Stern, S. A. Levin, M. A. Harwell, M. Alexander, and D. A. Andow. 1985. Potential Impacts of Environmental Release of Biotechnology Products: Assessment, Regulation, and Research Needs. Ithaca, NY.: Ecosystems Research Center, Cornell University, ERC-075. 219 pp.

Giovianni, S. J., E. F. Delong, G. J. Olsen, and N. R. Pace. 1988. Phylogenetic group-specific oligodeoxynucleotide probes for identification of single microbial cells. J. Bacteriol. 170:720-726.

Goldberg, R. B. 1988. Plants: Novel developmental processes. Science, 240:1460-1467.

Goldberg, R. B., S. J. Barker, and L. Perez-Grau. 1989. Regulation of gene expression during plant embryogenesis. Cell 56:149-160.

Golovleva, L. A., R. N. Pertsova, A. M. Boronin, V. M. Travkin, and S. A. Kozlovsky. 1988. Kelthane degradation by genetically engineered *Pseudomonas aeruginosa* BS827 in a soil ecosystem. Appl. Environ. Microbiol. 54:1587-1590.

Gottlieb, L. D. 1984. Genetics and morphological evolution in plants. Am. Nat. 123:681-709.

Graham, J. B., and C. A. Istock. 1979. Gene exchange and natural selection cause *B. subtilis* to evolve in soil culture. Science 204:637-639.

Greenberg, E. P., N. J. Poole, H. A. P. Pritchard, J. Tiedje, and D. E. Corpet. 1988. Use of microcosms. Pp. 266-274 in The Release of Genetically Engineered Micro-Organisms, M. Sussman, G. H. Collins, F. A. Skinner, and D. E. Stewart-Tall, eds. New York: Academic Press.

Grice, G., and M. R. Reeve, eds. 1982. Marine Mesocosms: Biological and Chemical Research in Experimental Ecosystems. New York: Springer-Verlag. 440 pp.

Grimes, D. J., and R. R. Colwell. 1986. Viability and virulence of *Escherichia coli* suspended by membrane chamber in semitropical ocean water. FEMS Microbiol. Lett. 34:161-165.

Grimes, D. J., R. W. Attwell, P. R. Brayton, L. M. Palmer, D. M. Rollins, D. B. Roszk, F. L. Singleton, M. L. Tamplin, and R. R. Colwell. 1986. The fate of enteric pathogenic bacteria in estuarine and marine environments. Microbiol. Sci. 3:324-329.

Grimsley, N., T. Hohn, J. W. Davies, and B. Hohn. 1987. *Agrobacterium*-mediated delivery of infectious maize streak virus into maize plants. Nature 325:177-179.

Guerry, P., and R. R. Colwell. 1977. Isolation of cryptic plasmid deoxyribonucleic acid from Kanagawa positive strains of *Vibrio parahaemolyticus*. Inform. Immunol. 16:328-334.

Ham, G. E. 1980. Inoculation of legumes with *Rhizobium* in competition with naturalized strains. Pp. 131-138 in Nitrogen Fixation, vol. 2, W. E. Newton and W. H. Orme-Johnson, eds. Baltimore: University Park Press.

Hammonds, A., ed. 1981. Ecotoxicological Test Systems. Washington, D.C.: Office of Toxic Substances, U.S. Environmental Protection Agency, ORNL-5709, EPA-560/11-80-004.

Harlan, J. R. 1975. Crops and Man. Madison, Wis.: American Society of Agronomy, Crop Science Society of America. 295 pp.

Harper, J. L. 1965. Establishment, aggression, and cohabitation in weedy species. Pp. 245-265 in The Genetics of Colonizing Species. H. G. Baker and G. L. Stebbins, eds. New York: Academic Press.

Harper, J. L. 1982. After description. Pp. 11-25 in The Plant Community as a Working Organism, Special publication of the British Ecological Society, E. I. Newman, ed. Oxford, England: Blackwell Scientific Publications.

Harte, J., D. Levy, J. Rees, and E. Saagebarth. 1980. Making microcosms an effective assessment tool. Pp. 105-107 in Microcosms in Ecological Research, J. P. Geisy, Jr., ed. Springfield, Va.: National Technical Information Service, DOE Symposium 52. 1008 pp.

Hartl, D. L., and D. E. Dykhuizen. 1984. The population genetics of *Escherichia coli*. Annu. Rev. Genet. 18:31-68.

Hartl, D. L., D. E. Dykhuizen, R. D. Miller, L. Green, and J. DeFramond. 1983. Transposable element IS*50* improves growth rate of *E. coli* cells without transposition. Cell 35:503-510.

Haughn, G., J. Smith, B. Mazur, and C. Somerville. 1988. Transformation with a mutant *Arabidopsis* acetolactate synthase gene renders tobacco resistant to sulfonylurea herbicides. Mol. Gen. Genet. 211:266-271.

Hawkes, J. G. 1982. History of the potato. Pp. 1-13 in The Potato Crop, P. M. Harris, ed. London: Chapman & Hall.

Heinemann, J. A., and G. F. Sprague, Jr. 1989. Bacterial conjugative plasmids mobilize DNA transfer between bacteria and yeast. Nature 340: 205-209.

Hemming, B. C., and D. J. Drahos. 1984. β-galactosidase, a selectable non-antibiotic marker for fluorescent pseudomonads. J. Cell. Biochem. Suppl. 8B:252.

Hershey, A. D., and R. Rotman. 1949. Genetic recombination between host range and plaque-type mutants of bacteriophage in single bacterial cells. Genetics 34:44-71.

Hinchee, M. A. W., D. V. Connorward, C. A. Newell, R. E. McDonnell, D. A. Fischhoff, D. B. Re, R. T. Fraley, and R. B. Horsch. 1988. Production of transgenic soybean using *Agrobacterium*-mediated DNA transfer. Biotechnology 6:915-921.

Hilder, V. A., A. M. R. Gatehouse, S. E. Sheerman, R. F. Barker, and D. Boulter. 1987. A novel mechanism of insect resistance engineered into tobacco. Nature 330:160-163.

Hill, D. L., T. J. Phelps, A. V. Palumbo, D. C. White, G. W. Strandberg, and T. L. Donaldson. In press. Bioremediation of polychlorinated biphenyls: Degradation capabilities in field lysimeters. Appl. Biochem. Biotechnol.

Hofte, H., and H. R. Whitely. 1989. Insecticidal crystal proteins of *Bacillus thuringiensis*. Microbiol. Rev. 53:242-255.

Holben, W. E., J. K. Jansson, B. K. Chelm, and J. M. Tiedje. 1988. DNA probe method for the detection of specific microorganisms in the soil bacterial community. Appl. Environ. Microbiol. 54:703-711.

Holm, L. G., D. L. Plucknett, J. V. Pancho, and J. P. Herberger. 1977. The World's Worst Weeds. Honolulu: The East-West Center, University of Hawaii.

Hulbert, L. C. 1955. Ecological studies of *Bromus tectorum* and other annual bromegrasses. Ecol. Monogr. 25:181-213.

Hunter, F. R., N. E. Cook, and P. F. Entwistle. 1984. Virus as pathogens for the control of insects. In Microbial Methods for Environmental Biotechnology, J. M. Grainger and J. M. Lynch, eds. New York: Academic Press.

Hymowitz, T. 1984. Dorsett-Morse soybean collection trip to East Asia: 50-year retrospective. Econ. Bot. 38:378-388.

Hymowitz, T. 1987. Introduction of soybean to Illinois. Econ. Bot. 41:28-32.

Hymowitz, T., and C. A. Newell. 1981. Taxonomy of the genus *Glycine*: Domestication and uses of soybeans. Econ. Bot. 35:272-288.

Infante, P. F., and T. A. Tsongas. 1982. Mutagenic and oncogenic effects of chloromethanes, chloroethanes, and halogenated analogs of vinyl chloride. Environ. Sci. Res. 25: 301-327.

Isensee, A. R., and N. Tayaputch. 1986. Distribution of carbofuran in a rice-paddy-fish microecosystem. Bull. Environ. Contam. Toxicol. 36:763-769.

Jain, S. K. 1977. Genetic diversity of weedy rye populations in California. Crop Sci. 17:480-482.

Jaynes, J. M., K. G. Xanthopoulos, L. Destefano-Beltran, and J. H. Dodds. 1987. Increasing bacterial disease resistance in plants utilizing antibacterial genes from insects. Bioessays 6:263-270.

Johnson, H. W., U. M. Mears, and C. R. Weber. 1965. Competition for nodule sites between strains of *Rhizobium japonicum*. Agron. J. 57:179-185.

Johnston, S. A., P. Q. Anziano, K. Shark, J. C. Sanford, and R. A. Butlow. 1988. Mitochondrial transformation in yeast by bombardment with microprojectiles. Science 240:1538-1541.

Jones, D. A., and A. Kerr. 1989. The efficacy of *Agrobacterium radiobacter* strain K1026, a genetically engineered derivative of strain K84, for the biological control of crown gall. Plant Dis. 73:15-18.

Jones, D. A., M. H. Ryder, B. G. Clare, S. K. Farrand, and A. Kerr. 1988. Construction of a Tra^- deletion mutant of pAgK84 to safeguard the biological control of crown gall. Mol. Gen. Genet. 212:207-214.

Keen, N. T., and B. Staskawicz. 1988. Host-range determinants in plant pathogens and symbionts. Annu. Rev. Microbiol. 42:421-440.

Kelman, A. 1953. The bacterial wilt caused by *Pseudomonas solanacearum*: A literature review and bibliography. N. Car. Exp. Sta. Tech. Bull. 99:194pp.

Kerr, A., and K. Htay. 1974. Biological control of crown gall through bacteriocin production. Physiol. Plant Pathol. 4:37-44.

Keyser, H. H., B. F. Weber, and S. L. Uratsu. 1984. *Rhizobium japonicum* serotype and hydrogenase phenotype distribution in 12 states. Appl. Environ. Microbiol. 47:613-615.

Kieft, T. L., E. Soroken, and M. K. Firestone. 1987. Microbial biomass response to a rapid increase in water potential when dry soil is wetted. Soil Biol. Biochem. 19:119-126.

Kilbane, J. J., D. K. Chatterjee, and A. M. Chakrabarty. 1983. Detoxification of 2,4,5-trichlorophenoxyacetic acid from contaminated soil by *Pseudomonas cepacia*. Appl. Environ. Microbiol. 45:1697-1700.

Kimbara, K., T. Hashimoto, M. Fukada, T. Koana, M. Takagi, M. Oishi, and K. Yano. 1989. Cloning and sequencing of two tandem genes involved in degradation of 2,3-dihydroxybiphenyl to benzoic acid in the polychlorinated biphenyl-degrading soil bacterium *Pseudomonas* sp. strain KKS102. J. Bacteriol. 171:2740-2747.

Klein, T., M. Fromm, A. Weissinger, D. Tomes, S. Schaaf, M. Sleetem, and J. Sanford. 1988. Transfer of foreign genes into intact maize cells using high-velocity microprojectiles. Proc. Natl. Acad. Sci. U.S.A. 85:4305-4309.

Knapp, S. J. 1988. Temperate industrial oilseed crops. Abstract in First National Symposium for New Crops: Research, Development, Economics, Indianapolis, Ind., October 23-26, 1988.

Kobayashi, H., and B. E. Rittman. 1982. Microbial removal of hazardous organic chemicals. Environ. Sci. Technol. 16:170A-181A.

Koncz, C., O. Olsson, W. H. R. Langridge, J. Schell, and A. A. Szalay. 1987. Expression and assembly of functional bacterial luciferase in plants. Proc. Natl. Acad. Sci. 84:131-135.

Konings, W. N., and E. H. Veldkamp. 1980. Phenotypic response to environmental change. Pp. 161-191 in Contemporary Microbiology, D. C. Elloond, J. N. Hedges, M. J. Lathan, J. M. Lynan, J. H. Slater, eds. London: Academic Press.

Konzak, C. F., S. A. Kleinhofs, and S. E. Ullrich. 1984. Induced mutations in seed-propagated crops. Plant Breeding Reviews 2:13-72.

Korwek, E. L. 1988. 1988 Biotechnology Regulations Handbook. Washington, D.C.: Center for Energy and Environmental Management.

Korwek, E. L., and P. L. de la Cruz. 1985. Federal regulation of environmental releases of genetically manipulated microorganisms. Rutgers J. Comput. Technol. Law. 118:801-886.

Kuhlemeier, C., P. J. Green, and N. H. Chua. 1987. Regulation of gene expression in higher plants. Annu. Rev. Plant Physiol. 38:221-257.

Kunkel, T. A. 1985. Rapid and efficient site-specific mutagenesis without phenotypic selection. Proc. Natl. Acad. Sci. U.S.A. 82:488-492.

Lee, S.W., and G. Edlin. 1985. Expression of tetracycline resistance in pBR322 derivatives reduces the reproductive fitness of plasmid-containing *Escherichia coli*. Gene 39:173-180.

Leffler, J. W. 1984. The use of self-selected, generic aquatic microcosms for pollution effects assessment. Pp. 139-157 in Concepts in Marine Pollution Measurements, H. H. White, ed. College Park, Md.: Maryland Sea Grant College, University of Maryland.

Lenski, R. E. 1987. The infectious spread of engineered genes. Pp. 99-124 in Application of Biotechnology: Environmental and Policy Issues, J. R. Fowle III, ed. Washington, D.C.: American Association for the Advancement of Science.

Lenski, R. E. In press. Fitness and Gene Stability, M. Levin, R. Seidler, P. H. Pritchard, and M. Rogul, eds. Washington, D.C.: American Society for Microbiology.

Lenski, R. E., and T. T. Nguyen. 1988. Stability of recombinant DNA and its effects on fitness. Pp. 18-20 in Planned Release of Genetically Engineered Organisms (Special Issue of Trends in Biotechnology and Trends in Ecology and Evolution), J. Hodgson and A. M. Sugden, eds. Cambridge, England: Elsevier.

Lenski, R. E., and S. E. Hattingh. 1986. Coexistence of two competitors on one resource and one inhibitor: A chemostat model based on bacteria and antibiotics. J. Theor. Biol. 122:83-93.

Levin, B. R. 1984. Changing views of the hazards of recombinant DNA manipulation and the regulation of these procedures. Recomb. DNA Tech. Bull. 7:107-114.

Levin, B. R., and V. A. Rice. 1980. The kinetics of transfer of nonconjugative plasmids by mobilizing conjugative factors. Genet. Res. Cambridge 35:241-259.

Levy, S. B., and R. P. Novick, eds. 1986. Antibiotic Resistance Genes: Ecology, Transfer, and Expression. Cold Spring Harbor, N.Y.: Cold Spring Harbor Laboratory.

Liang, L. N., J. L. Sinclair, L. M. Mallory, and M. Alexander. 1982. Fate in model ecosystems of microbial species of potential use in genetic engineering. Appl. Environ. Microbiol. 44:708-714.

Lighthart, B., Baham, J., and V. V. Volk. 1982. Microbial respiration and chemical speciation in metal-amended soils. J. Environ. Qual. 12:543-548.

Lindow, S. E. In press. Construction of isogenic ice-strains of *Pseudomonas syringae* for evaluation of specificity of competition on leaf surfaces. In Microbial Ecology, F. Megusar, ed.

Lindow, S. E., and N.J. Panopoulos. 1988. Field tests of recombinant ice-*Pseudomonas syringae* for biological frost control in potato. Pp. 121-138 in The Release of Genetically Engineered Micro-Organisms, M. Sussman, G. H. Collins, F. A. Skinner, and D. E. Stewart-Tall, eds. London: Academic Press.

Lindow, S. E., N. J. Panopoulos, and B. L. McFarland. 1989. Genetic engineering of bacteria from managed and natural habitats. Science 244:1300-1307.

Little, C. D., A. V. Palumbo, S. E. Herbes, M. E. Lidstrom, R. L. Tyndall, and P. J. Gilmer. 1988. Trichloroethylene biodegradation by a methane-oxidizing bacterium. Appl. Environ. Microbiol. 54: 951-956.

Livingston, R. J., R. J. Diaz, and D. C. White. 1985. Field validation of laboratory-derived multispecies aquatic test systems. Gulf Breeze, Fla.: U.S. Environmental Protection Agency, Environmental Research Laboratory, EPA/600/S4-85/039.

Mach, P. A., and D. J. Grimes. 1982. R-plasmid transfer in a wastewater treatment plant. Appl. Environ. Microbiol. 44:1395-1403.

Mack, R. N. 1981. Invasion of *Bromus tectorum* L. into western North America: An ecological chronicle. Agro-Ecosystems 7:145-165.

Mack, R. N. 1985. Invading plants: Their potential contribution to population biology. Pp. 127-142 in Plant Populations: Essays in Honour of John L. Harper, J. White, ed. London: Academic Press.

Mazodier, P., R. Petter, and C. Thompson. 1989. Intergeneric conjugation between *Escherichia coli* and *Streptomyces* species. J. Bacteriol. 171:3583-3585.

Mack, R. N. 1986. Alien plant invasions into the intermountain west: A case history. Pp. 191-213 in Ecology of Biological Invasions of North America and Hawaii. Ecological Studies: Vol 58. H. A. Mooney and J. A. Drake, eds. New York: Springer-Verlag.

McCabe, D. E., W. F. Swain, B. J. Martinell, and P. Christou. 1988. Stable transformation of soybean glycine-max by particle acceleration. Bio/Technology 6:923-926.

McClintock, B. 1950. Mutable loci in maize. Carnegie Inst. Washington Yearb. 49:157-167.

McConnell, M. M., H. R. Smith, J. A. Willshaw, A. M. Field, and B. Rowe. 1981. Plasmids coding for colonization factor antigen I and heat-stable enterotoxin production isolated from enterotoxigenic *Escherichia coli*: Comparison of their profiles. Infect. Immunol. 32:927-36.

McCormick, D. 1985. No escaping free release. Bio/Technology 3:1065-1067.

McFeters, G. A., and D. G. Stuart. 1972. Survival of coliform bacteria in natural waters: Field and laboratory studies with membrane filter chambers. Appl. Environ. Microbiol. 24:805-811.

McGarity, T. O., and K. O. Bayer. 1983. Federal regulation of emerging genetic technologies. Vanderbilt Law Rev. 36:461-540.

Mellano, V. J., and D. A. Cooksey. 1978. Development of host-range mutants of *Xanthomonas campestris* pv. *translucens*. Appl. Environ. Microbiol. 54:884-889.

Metcalf, R. L., G. K. Sangha, and I. P. Kapoor. 1971. Model ecosystem for the evaluation of pesticide biodegradability and ecological magnification. Environ. Sci. Technol. 5:709-713.

Miller, J. F., J. K. Mekalanos, and S. Falkow. 1989. Coordinate regulation and sensory transduction in the control of bacterial virulence. Science 243:916-922.

Miller, R. V. 1988. Potential for gene transfer and establishment of engineered genetic sequences. Pp. 23-27 in Planned Release of Genetically Engineered Organisms (Special Issue of Trends in Biotechnology/Trends in Ecology and Evolution), J. Hodgson and A. M. Sugden, eds. Cambridge, England: Elsevier.

Miller, J. H. 1983. Kudzu: Where did it come from? And how do we stop it? Southern J. Appl. Forestry 7:165-169.

Mims, C. A. 1982. The Pathogenesis of Infectious Disease. New York: Academic Press. 297 pp.

Moawad, H. W., W. R. Ellis, and E. L. Schmidt. 1984. Rhizosphere response as a factor in competition among three serogroups of indigenous *Rhizobium japonicum* for nodulation of field-grown soybeans. Appl. Environ. Microbiol. 47:607-612.

Molin, S. P., P. Klemm, L. K. Poulsen, H. Biehl, K. Gerdes, and P. Andersson. 1987. Conditional suicide system for containment of bacteria and plasmids. Bio/Technology 5:1315-1318.

Morese, S. A., S. R. Johnson, J. W. Biddle, and M. C. Roberts. 1986. High-level tetracycline resistance in *Neisseria gonorrhoeae* is result of acquisition of streptococcal tet M determinant. A.A.C. 30:664-670.

Morita, R. Y. 1982. Starvation-survival of heterotrophs in the marine environment. Adv. Microbial. Ecol. 6:171-198.

Morton, J. F. 1978. Brazilian pepper—Its impact on people, animals and the environment. Econ. Bot. 32:353-359.

Morton, J. F. 1980. The Australian pine or beefwood (*Casuarina equisetifolia*), an invasive "weed" tree in Florida. Proc. Fla. State Hort. Soc. 93:87-95.

NAS (National Academy of Sciences). 1977. Research with Recombinant DNA: An Academy Forum. Washington, D.C.: National Academy of Sciences. 52 pp.

NAS (National Academy of Sciences). 1987. Introduction of Recombinant DNA-engineered Organisms into the Environment: Key Issues. Washington, D.C.: National Academy Press. 24 pp.

Nelson, M. J. K., S. O. Montgomery, E. J. O'Neill, and P. H. Pritchard. 1986. Aerobic metabolism of trichloroethylene by a bacterial isolate. Appl. Environ. Microbiol. 52:383-384.

Nelson, R., S. McCormick, X. Delannay, P. Dube, J. Layton, E. Anderson, M. Kaniewski, R. Proksch, R. Horsch, S. Rogers, R. Fraley, and R. Beachy. 1988. Virus tolerance, plant growth, and field performance of transgenic tomato plants expressing coat protein from tobacco mosaic virus. Bio/Technology 6:1394-1395.

Nester, E. W., M. P. Gordon, R. M. Amasino, and M. F. Yanofsky. 1984. Crown gall: A molecular and physiological analysis. Annu. Rev. Plant Physiol. 35:387-413.

Nester, E. W., and T. Kosuge. 1981. Plasmids specifying plant hyperplasias. Annu. Rev. Microbiol. 35:531-565.

New, P. B., and A. Kerr. 1972. Biological control of crown gall: Field measurements and glasshouse experiments. J. Appl. Bacteriol. 35:279-287.

Nicolaides, A. A. 1987. Microbial mineral processing: The opportunities for genetic manipulation. J. Chem. Tech. Biotechnol. 38:167-185.

NIH (National Institutes of Health). 1976. Recombinant DNA research: Guidelines. Fed. Reg. 41(July 7):27902.

NIH (National Institutes of Health). 1978. Guidelines for research involving recombinant DNA molecules. Fed. Reg. 43(December 22):60108.

NIH (National Institutes of Health). 1981. Recombinant DNA research: Action under guidelines. Fed. Reg. 46(August 7):40331.

NIH (National Institutes of Health). 1983. Recombinant DNA research: Action under guidelines. Fed. Reg. 48(April 15):16459.

NIH (National Institutes of Health). 1983. Recombinant DNA research: Actions under guidelines. Fed. Reg. 48(June 1):24548.

Nilson, E. B., D. L. Devlin, and D. G. Mosier. 1988. Shattercane Control in Field Crops. North Central Regional Extension Publ. 297. 11pp.

Nutman, P. S. 1975. Rhizobium in the soil. Pp. 111-131 in Soil Microbiology, N. Walker, ed. New York: Academic Press.

Obukowicz, M. G., F. J. Perlak, K. Kusano-Kretzmer, E. J. Meyer, S. L. Bolten, and L. S. Watrud. 1987. IS*50*L as a non-self transposable vector used to integrate the *Bacillus thuringiensis* delta-endotoxin gene into the chromosome of root-colonizing Pseudomonads. Gene 51:91-96.

Ogram, A. V., and G. S. Sayler. 1988. The use of gene probes in the rapid analysis of natural microbial communities. J. Indust. Microbiol. 3:281-292.

Olson, B. H., and T. Barkay. 1986. Feasibility of using bacterial resistance to metals in mineral exploration. Pp. 311-327 in Mineral Exploration: Biological Systems and Organic Matter, D. Carlisle, W. Berry, I. Kaplan, and J. Watterson, eds. Englewood Cliffs, N.J.: Prentice-Hall.

Olson, B. H., and I. Thornton. 1982. The development of a bacterial indicator system to assess bioavailability of metals in contaminated land. J. Soil Sci. 33:271-279.

Olson, S. 1986. The molecular and microbial products of biotechnology. Pp. 14-29 in Biotechnology: An Industry Comes of Age. Washington, D.C.: National Academy Press.

Ooms, G., M. M. Burrell, A. Karp, M. Bevan, and J. Hille. 1987. Genetic transformation in two potato cultivars with T-DNA from disarmed *Agrobacterium*. Theor. Appl. Genet. 73:744-750.

Orrego, C., A. Arnaudo, and H. O. Halvorson. 1978. *Bacillus subtilus* 168 genetic transformation mediated by outgrowing spores: Necessity for cell contact. J. Bacteriol. 134:973-981.

Orser, C. S., R. Lotstein, B. J. Staskawicz, D. Dahlbeck, E. Lahue, D. K. Willis, S. E. Lindow, and N. J. Panopoulos. 1984. Molecular genetics of bacterial ice nucleation. Pp. 98-107 in Proceeding of the Second Working Group on *Pseudomonas syringae pathovars*, C. G. Panagopoulos, P. G. Psallidas, and A. S. Alivizatos, eds. Athens: Hellenic Phytopathological Society.

Orton, T. J. 1983. Somaclonal variation: Theoretical and practical considerations. Pp 427-468 in Gene Manipulations in Plant Improvement, J. P. Gustafson, ed. New York: Plenum.

OSTP (Office of Science and Technology Policy). 1984. Proposal for a coordinated framework for regulation of biotechnology. Fed. Reg. 49(December 31):50856.

OSTP (Office of Science and Technology Policy). 1985. Coordinated framework for regulation of biotechnology: Establishment of the Biotechnology Science Coordinating Committee. Fed. Reg. 50(November 14):47174.

OSTP (Office of Science and Technology Policy). 1986. Coordinated framework for regulation of biotechnology. Fed. Reg. 51(June 26):23302.

OTA (Office of Technology Assessment). U.S. Congress. 1984. Commercial Biotechnology: An International Analysis. DC OTA-BA-218, 160 pp. Washington, D.C.: Government Printing Office.

OTA (Office of Technology Assessment). U.S. Congress. 1988. New Developments in Biotechnology Field-Testing Engineered Organisms: Genetic and Ecological Issues. OTA-BA-350. Washington, D.C.: Government Printing Office. 150 pp.

Ow, D. W., K. V. Wood, M. Deluca, J. R. DeWet, D. R. Helinski, and S. H. Howell. 1986. Transient and stable expression of the firefly luciferase gene in plant cells and transgenic plants. Science 234:856-859.

Panagopoulos, C. G., P. G. Psallidas, and A. S. Alivijatos. 1979. Evidence of a breakdown in the effectiveness of biological control of crown gall. Pp. 569-578 in Soil-Borne Plant Pathogens, B. Schippers and W. Garns, eds. London: Academic Press.

Parker, C., and M. L. Dean. 1976. Control of wild rice in rice. Pestic. Sci. 7:403-416.

Parker, C. A., M. J. Trinick, and D. L. Chatel. 1977. *Rhizobia* as soil and rhizosphere inhabitants. Pp. 311-352 in A Treatise on Dinitrogen Fixation, R. W. F. Hardy and A. H. Gibson, eds. New York: Wiley-Interscience.

Paszkowski, J., M. Baur, A. Bogucki, and I. Potrykus. 1988. Gene targeting in plants. EMBO J. 7:4021-4026.

Peerbolte, R., K. Leenhouts, G. M. S. Hooykaas-Van Slogteren, J. Hoge, G. J. Wullems, and R. A. Schillperoon. 1986. Clones from a shooty tobacco crown gall tumor. I. Deletions, rearrangements and amplifications resulting in irregular T-DNA structures and organizations. Plant Mol. Biol. 7:265-284.

Peters, N. K., J. W. Frost, and S. R. Long. 1986. A plant flavone, luteolin, induces expression of *Rhizobium meliloti* nodulation genes. Science 233:977-980.

Pickersgill, B. 1981. Biosystematics of crop-weed complexes. Kulturpflanze 29:377-388.

Podgewaite, J. D. 1985. Strategies for field use of baculoviruses. Pp. 775-797 in Viral Insecticides for Biological Control, K. Maramorosch and K. E. Sherman, eds. New York: Academic Press.

Powell-Abel, P., R. S. Nelson, B. De, N. Hoffmann, S. G. Rogers, R. T. Fraley, and R. N. Beachy. 1986. Delay of disease development in transgenic plants that express the tobacco mosaic virus coat protein gene. Science 232:738-743.

Prentki, P., and H. M. Krisch. 1984. In vitro insertional mutagenesis with a selectable DNA fragment. Gene 29:303-313.

Pritchard, P. H., and A. W. Bourquin. 1984. The use of microcosms for evaluation of interactions between pollutants and microorganisms. Adv. Microb. Ecol. 7:133-215.

Rake, T. M., C. Hagedorn, E. L. McCoy, and G. F. Klug. 1978. Transport of antibiotic-resistant *Escherichia coli* through western Oregon hillslope soils under conditions of saturated flow. J. Environ. Qual. 7:487-494.

Quensen, J. F., III, J. M. Tiedje, and S. A. Boyd. 1988. Reductive dechlorination of polychlorinated biphenyls by anaerobic microorganisms from sediments. Science 242:752-754.

Reddy, P. M., and Roger, P. A. 1988. Dynamics of algal populations and acetylene-reducing activity in five rice soils inoculated with blue-green algae. Biol. Fertil. Soils 6:14-21.

Redmond, J. W., M. Bately, M. A. Djordjevic, R. W. Innes, P. L. Kuempel, and B. G. Rolfe. 1986. Flavones induce expression of nodulation genes in *Rhizobium*. Nature 323:632-634.

Reyes, V. G., and J. M. Tiedje. 1976. Ecology of the gut microbiota *Tracheoniseus rathlesi*. Pedobiologia 16:67-74.

Reznikoff, W., and L. Gold. 1985. Maximizing Gene Expression. Boston: Butterworth Publishers. 375 pp.

Rhodes, C. A., D. A. Pierce, I. J. Metler, D. Mascarenhas, and J. Detmer. 1988. Genetically transformed maize plants from electroporated protoplasts. Science 240:204-207.

Rieseberg, L. H., D. E. Soltis, and J. D. Palmer. 1988. A molecular reexamination of introgression between *Helianthus annus* and *H. bolanderi* (*Compositae*). Evolution 42:227-238.

Roberts, M., L. P. Elwell, and S. Falkow. 1977. Molecular characterization of two beta-lactamase-specifying plasmids isolated from *Neisseria gonorrhoeae*. J. Bacteriol. 131:557-563.

Rochkind-Dubinsky, M. L., G. S. Saylor, and J. W. Blackburn. 1987. Microbiological Decomposition of Chlorinated Aromatic Compounds. New York: Dekker.

Ross, A. 1986. Potato Breeding: Problems and Perspectives. Berlin: Parey. 132 pp.

Roszak, D. B., and R. R. Colwell. 1987. Survival strategies of bacteria in the natural environment. Microbiol. Rev. 51:365-379.

Rottmann, W. H., T. Brears, T. P. Hodge, and D. M. Longdale. 1987. A mitochondrial gene is lost via homologous recombination during reversion of CMS T maize to fertility. EMBO J. 6:1541-1546.

Rouse, D. I., Nordheim, E. V., Hirano, S. S., and C. D. Upper. 1985. A model relating the probability of foliar disease incidence to the population frequencies of bacterial plant pathogens. Phytopathology 75:505-509.

Ruvken, G. B., and F. M. Ausubel. 1981. A general method for site-directed mutagenesis in prokaryotes. Nature 289:85-88.

Ryan, C. A. 1988. Proteinase inhibitor gene families: Tissue specificity and regulation. Pp. 223-233 in Temporal and Spatial Regulation of Plant Genes, D. P. S. Verma and R. B. Goldberg, eds. Vienna: Springer-Verlag.

Salisbury, E. J. 1961. Weeds and Aliens. London: Collins.

Sangodkar, U. M. X., P. J. Chapman, and A. M. Chakrabarty. 1988. Cloning, physical mapping, and expression of chromosomal genes specifying degradation of the herbicide 2,4,5-T by *Pseudomonas cepacia* ACllOO. Gene 71:267-277.

Sauer, J. D. 1967. The grain amaranths and their relatives: A revised taxonomic and geographic survey. Ann. Mo. Bot. Gard. 54:103-137.

Scanferlato, U. S., D. R. Orvos, J. Cairns, and G. H. Lary. 1989. Genetically engineered *Erwinia carotovora* in aquatic microcosms: Survival and effects on functional groups of indigenous bacteria. Appl. Environ. Microbiol. 55:1477-1482.

Schmidt, E. L., and F. M. Robert. 1985. Recent advances in the Ecology of Rhizobium. Pp. 279-385 in Nitrogen Fixation Research Progress. 6th International Symposium on Nitrogen Fixation, H. J. Evans, P. J. Bottomley, and W. E. Newton, eds. Dordrecht, Netherlands: Martinus Nijhoff Publishers.

Schofield, P. R., A. H. Gibson, W. F. Dudman, and J. T. Watson. 1987. Evidence for genetic exchange and recombination of *Rhizobium* symbiotic plasmids in a soil population. Appl. Environ. Microbiol. 53:2942-2947.

Scolnik, P. A., and R. Haselkorn. 1984. Activation of extra copies of genes coding for nitrogenase in *Rhodopseudomonas capsulata*. Nature 307:289-292.

Second, G. 1982. Origin of the genetic diversity of cultivated rice (*Oryza* spp.): Study of the polymorphism scored at 40 isozyme loci. Jpn. J. Genet. 57:25-57.

Sengupta-Gopalan, C., N. A. Reichert, R. F. Barker, T. C. Hall, and J. D. Kemp. 1985. Developmentally regulated expression of the bean-phaseolin gene in tobacco seed. Proc. Natl. Acad. Sci. U.S.A. 82:3320-3324.

Senior, E., A. T. Bull, J. H. Slater. 1976. Enzyme evolution in a microbial community growing on the herbicide Dalapon. Nature 263:476-479.

Sequeira, L. 1984. Plant-bacterial interactions. Pp. 187-211 in Cellular Interactions, H. F. Linskens and J. Heslop-Harrison, eds. Berlin: Springer-Verlag.

Shah, D. M., R. B. Horsch, H. J. Klee, G. M. Kishore, J. A. Winter, N. E. Tumer, C. M. Hironaka, P. R. Sander, C. S. Gasser, S. Aykent, N. R. Siegel, S. G. Rogers, and R. T. Fraley. 1986. Engineering herbicide tolerance in transgenic plants. Science 233:478-481.

Sheehy, R. E., M. Kramer, and W. R. Hiatt. 1988. Reduction of polygalacturonase in tomato fruit by antisense RNA. Proc. Natl. Acad. Sci. U.S.A. 85:8805-8809.

Shields, M. S., S. W. Hooper, and G. S. Saylor. 1985. Plasmid-mediated mineralization of 4-chlorobiphenyl. J. Bacteriol. 163:882-889.

Simberloff, D. 1985. Predicting ecological effects of novel entities: evidence from higher organisms. Pp. 152-161 in Engineered Organisms in the Environment: Scientific Issues, H. O. Halvorson, D. Pramer, and M. Rogul, eds. Washington D.C.: American Society for Microbiology

Simmonds, N. W., ed. 1979. Evolution of Crop Plants. New York: Longman. 339 pp.

Slota, J. E., and S. K. Farrand. 1982. Genetic isolation and physical characterization of pAgK84, the plasmid responsible for agrocin 84 production. Plasmid 8:175-186.

Small, E. 1984. Hybridization in the domesticated-weed-wild complex. Pp. 195-210 in Plant Biosystematics, W. F. Grant, ed. Ontario: Academic Press.

Smith, C. W. 1985. Impact of alien plants on Hawaii's native biota. Proceedings of the Fourth National Park Science Conference. Cooperative Park Study Unit, Honolulu, Hawaii.

Smith, J. S. C., M. M. Goodman, and R. N. Lester. 1981. Variation within teosinte. I. Numerical analysis of morphological data. Econ. Bot. 3535:187-203.

Smith, M. S., G. W. Thomas, R. E. White, and R. Ritonga. 1985. Transport of *Escherichia coli* through intack and disturbed soil columns. J. Environ. Qual. 14:87-91.

Somerville, C. C., I. T. Knight, W. C. Straube, and R. R. Colwell. 1989. Simple, rapid method for direct isolation of nucleic acids from aquatic environments. Appl. Environ. Microbiol. 55:548-554.

Stachel, S. E., E. Messens, M. Van Montagu, and P. Zambryski. 1985. Identification of the signal molecules produced by wounded plants that activate T-DNA transfer in *Agrobactertium tumefaciens*. Nature 318:624-629.

Stachel, S. E., E. W. Nester, and P. C. Zambryski. 1986. A plant cell factor induces *Agrobacterium tumefaciens vir* gene expression. Proc. Natl. Acad. Sci. U.S.A. 83:379-383.

Stalker, D., K. McBride, and L. Malyj. 1988. Herbicide resistance in transgenic plants expressing a bacterial detoxification gene. Science 242:419-422.

Staskawicz, B., D. Dahlbeck, and N. T. Keen. 1984. Cloned avirulence gene of *Pseudomonas syringae* pv. *glycinea* determines race-specificity incompatibility on *Glycine max* (L.) Merr. Proc. Natl. Acad. Sci. U.S.A. 81:6024-6028.

Staskawicz, B., U. Bonas, D. Dahlbeck, T. Huynk, B. Kearney, P. Ronald, and M. Whalen. 1988. Molecular determinants of specificity in plant bacterial interactions. Pp. 124-130 in Physiology and Biochemistry of Plant-Microbial Interactions, N. T. Keen, T. Kosuge, and L. Walling, eds. Rockville, Md.: American Society of Plant Physiologists.

Steffan, R. J., and Atlas, R. M. 1988. DNA amplification to enhance detection of genetically engineered bacteria in environmental samples. Appl. Environ. Microbiol. 54: 2185-2191.

Steffan, R. J., J. Goksoyr, A. K. Bej, and R. M. Atlas. 1988. Recovery of DNA from soils and sediments. Appl. Environ. Microbiol. 54:2908-2915.

Stetzenbach, L. D., B. Lighthart, R. J. Seidler, and S. C. Hern. 1989. Factors influencing the dispersal and survival of aerosolized microorganisms.

Stotzky, G., and H. Babich. 1986. Survival of, and genetic transfer by, genetically engineered bacteria in natural environments. Adv. Appl. Microbiol., xxx: 93-138.

Strauss, H., D. Hattis, and G. Page. 1986. Genetically Engineered Microorganisms: II. Survival Multiplication and Genetic Transfer. Recomb. DNA Tech. Bull. 9:69-88.

Suneson, C. A., K. O. Rachie, and G. S. Khush. 1969. A dynamic population of weedy rye. Crop Sci. 9:121-124.

Taira, K, N. Hayase, N. Arimura, S. Yamashida, D. Miyazaki, and K. Furukawa. 1988. Cloning the nucleotide sequence of the 2, 3-dihydroxybiphenyl dioxygenase gene from the PCB degrading strain of *Pseudomonas paucimobilies* Q1. Biochemistry 27:3990-3996.

Tanksley, S. D., N. D. Young, A. H. Peterson, and M. W. Bonierbale. 1989. RFLP mapping in plant-breeding: New tools for an old science. Bio/Technology 7:257-264.

Taub, F. B., and M. E. Crow. 1980. Synthesizing aquatic microcosms. Pp. 69-104 in Microcosms in Ecological Research, J. P. Geisy, Jr., ed. Springfield, Va.: National Technical Information Service, DOE Symposium No. 52.

Taylor, J. W., J. Oh, and F. Eckstein. 1985. The rapid generation of oligonucleotide-directed mutations at high frequency using phosphorothioate-modified DNA. Nucl. Acid Res. 13:8765-8785.

Tempé, J., and A. Petit. 1983. La pist des opinies. Pp. 14-32 in Molecular Genetics of the Bacteria-Plant Interaction, A. Puhler, ed. Berlin: Springer-Verlag.

Thomsen, C. D., G. D. Barbe, W. A. Williams, and M. R. George. 1986. "Escaped" artichokes are troublesome pests. Calif. Agric. 40:7-9.

Tiedje, J. M. 1987. Environmental monitoring of microorganisms. Pp. 115-127 in Prospects for Physical and Biological Containment of Genetically Engineered Organisms, J. W. Gillett, ed. Ithaca, N.Y.: Ecosystems Research Center Report. 114, Cornell University.

Tiedje, J. M., R. K. Colwell, Y. L. Grossman, R. Hodson, R. E. Lenski, R. N. Mack, and P. J. Regal. 1989. The planned introduction of genetically engineered organisms: Ecological considerations and recommendations. Ecology 70:298-315.

Timmis, K. M., G. J. Warren, H. R. Whitely, N. Gutterson, and M. J. Gasson. Round Table 1: Applications. Pp. 193-206 in The Release of Genetically Engineered Micro-Organisms, M. Sussman, G. H. Collins, F. A. Skinner, and D. E. Stewart-Tall, eds. New York: Academic Press.

Tolin, S., and A. K. Vidaver. 1989. Guidelines and regulations for research with genetically modified organisms: a view from academe. Annu. Rev. Phytopathol 27:551-581.

Tomasek, P. H., B. Frantz, U. M. X. Sangodkar, R. A. Haugland, and A. M. Chakrabarty. 1989. Characterization and nucleotide sequence determination of a repeated element isolated from a 2,4,5-T degrading strain of *Pseudomonas cepacia*. Gene 76:227-238.

Trevors, J. T., T. Barkay, and A. W. Bourquin. 1987. Gene transfer among bacteria in soil and aquatic environments: A review. Can. J. Microbiol. 33:191-198.

Tucker, J. M., and J. D. Sauer. 1958. Aberrant Amaranthus populations of the Sacramento-San Joaquin delta, California. Madrono 14:252-261.

Tumer, N. E., K. M. O'Connell, R. S. Nelson, P. R. Sanders, R. N. Beachy, R. T. Fraley, and D. M. Shah. 1987. Expression of alfalfa mosaic virus coat protein gene confers cross-protection in transgenic tobacco and tomato plants. EMBO J. 6:1181-1188.

Unterman, R., D. L. Bedard, M. J. Brennan, L. H. Bopp, J. J. Mondello, R. E. Brooks, D. P. Mobley, J. B. McDermott, C. C. Schwartz, and D. K. Diedrich. 1987. Biological approaches for polychlorinated biphenyl degradation. Pp. 253-269 in Environmental Biotechnology: Reducing Risk from Environmental Chemicals Through Biotechnology, G. S. Omenn, ed. New York: Plenum.

U.S. Congress. 447 U.S.C. 303 (1980).

U.S. Congress. 42 U.S.C. 4321-4347 (1982).

U.S. Congress. 1983. Environmental implications of genetic engineering. 98th Cong., 1st sess. Hearing Before the Subcommittee on Investigations and Oversight and the Subcommittee on Science, Research and Technology. Committee on Science and Technology. Washington, D.C.: Government Printing Office.

USDA (U.S. Department of Agriculture). 1986. Handbook of Agricultural Statistical Data. Washington, D.C.: Government Printing Office. 139 pp.

U.S. Environmental Protection Agency. 1985. Substances found at proposed and final NPL sites through update number three. Washington, D.C.: U.S. Environmental Protection Agency, Document NPL-U3-6-3.

U.S. Senate. 1984. The potential environmental consequences of genetic engineering. 98th Cong., 2d sess. September 25 and 27. Hearings Before the Subcommittee on Toxic Substances and Environmental Oversight. Committee on Environment and Public Works. Washington, D.C.: Government Printing Office.

Vaeck, M., A. Reynaerts, H. Hofte, S. Jansens, M. M. De Beuckeleer, C. Dean, M. Zabeau, M. Van Montagu, and J. Leemans. 1987. Transgenic plants protected from insect attack. Nature 328:33-37.

Van Voris, P., R. V. O'Neill, W. R. Emanuel, and H. H. Shugart. 1980. Functional complexity and ecosystem stability. Ecology 6:1352-1360.

Van Voris, P., D. A. Tolle, M. F. Arthur, J. Chesson, and R. W. Brocksen. 1983. Terrestrial microcosms: Validation, applications, and cost-benefit analysis. Paper presented to the Society of Environmental Toxicology and Chemistry, Arlington, Va.

Van der Krol, A. R., P. E. Lenting, J. Veenstra, I. M. Van der Meer, R. E. Koes, A. G. M. Gerats, J. N. M. Mol, and A. R. Stuitje. 1988. An anti-sense chalcone synthase gene in transgenic plants inhibits flower pigmentation. Nature 333:866-869.

Vandenbergh, M. P. 1986. The rutabaga that ate Pittsburgh: Federal regulation of free release biotechnology. Va. Law Rev. 72:1529-1568.

Vasil, I. K. 1988. Progress in the regeneration and genetic manipulation of cereal crops. Bio/Technology 6:397-402.

Vasil, I. K., ed. 1986. Cell Cultures and Somatic Cell Genetics of Plants, 3. Plant Regeneration and Genetic Variability. New York: Academic Press. 657 pp.

Vidaver, A., and G. Stotzky. In preparation. Decontamination and mitigation.

Vitousek, P. M. 1986. Biological invasions and ecosystem properties. Can species make a difference? Pp. 163-178 in Ecology of Biological Invasions in North America and Hawaii, H. A. Mooney and J. A. Drake, eds. New York: Springer-Verlag.

Vogel, T. M., and P. L. McCarty. 1985. Biotransformation of tetrachloroethylene to trichloroethylene, dichloroethylene, vinyl chloride, and carbon dioxide under methanogenic conditions. Appl. Environ. Microbiol. 49:1080-1083.

Wallroth, M., A. G. M. Gerats, S. G. Rogers, R. T. Fraley, and R. B. Horsch. 1986. Chromosomal localization of foreign genes in *Petunia hybrida*. Mol. Gen. Genet. 202:6-15.

Warwick, S. I., and L. D. Black. 1983. The biology of Canadian weeds. 61. *Sorghum halepense* (L.) PERS. Can. J. Plant Sci. 63:997-1014.

Warwick, S. I., B. K. Thompson, and L. D. Black. 1984. Population variation in *Sorghum halepense*, Johnson grass, at the northern limits of its range. Can. J. Bot. 62:1781-1790.

Wilson, J. T., J. F. McNabb, B. H. Wilson, and M. J. Noonan. 1983. Biotransformation of selected organic pollutants in ground water. Dev. Ind. Microbiol. 24:225-233.

Windle, P. N., and E. H. Franz. 1979a. Plant population structure and aphid parasitism: Changes in barley monocultures and mixtures. J. Appl. Ecol. 16:259-268.

Windle, P. N., and E. H. Franz. 1979b. The effects of insect parasitism on plant competition: Greenbugs and barley. Ecology 60:521-529.

Winter, R. B., K. Yen, and B. Ensley. 1989. Efficient degradation of trichloroethylene by a recombinant *Escherichia coli*. Bio/Technology 7:282-285.

Wong, W. K., C. Curry, R. S. Parekh, S. R. Parekh, M. Wayman, R. W. Davies, D. G. Kilburn, and N. Skipper. 1988. Wood hydrolysis by *Cellulomonas fimi* endoglucanase and exoglucanase coexpressed as secreted enzymes in *Saccharomyces cerevisiae*. Bio/Technology. 6:713-719.

Wright, S. F., J. G. Foster, and O. L. Bennett. 1986. Production and use of monoclonal antibodies for identification of strains of *Rhizobium trifolii*. Appl. Environ. Microbiol. 52:119-123.

Yates, J. R., R. P. Cunningham, and D. S. Holmes. 1988. IST2: An insertion sequence from *Thiobacillus ferrooxidans*. Proc. Natl. Acad. Sci. U.S.A. 85:7284-7287.

Yoder, 0. C. 1980. Toxins in pathogenesis. Annu. Rev. Phytopathol. 18:103-129.

Zadoks, J., and R. O. Schein. 1979. Epidemiology and Plant Disease Management. New York: Oxford University Press. 427 pp.

Zelibor, J. L., Jr., M. W. Doughten, D. H. Grimes, and R. R. Colwell. 1987. Testing for bacterial resistance to arsenic in monitoring well water samples by the direct viable counting method. Appl. Environ. Microbiol. 53:2929-2934.

Zund, P., and G. Lebek. 1980. Generation time-prolonging R plasmids: Correlation between increases in the generation time of *Escherichia coli* caused by R plasmids and their molecular size. Plasmid 3:65-69.

Information on Committee Members

STEERING COMMITTEE

Robert H. Burris (Chairman) is an emeritus professor of biochemistry at the University of Wisconsin, Madison. His scientific accomplishments include research on biological nitrogen fixation, respiration of plants, photosynthesis, and hydrobiology. He has served on numerous committees of the NAS/NRC and on national and international panels. He is a member of the National Academy of Sciences, the American Academy of Arts and Sciences, and the American Philosophical Society. He is a past president of the American Society of Plant Physiologists.

Fakhri A. Bazzaz is the H. H. Timken Professor of Science in the Department of Organismic and Evolutionary Biology, Harvard University. His research includes physiological ecology, plant community organization, the recovery of damaged ecosystems, and the rising CO_2 concentrations and global change. He has served on several NAS/NRC committees and is a member of the American Academy of Arts and Sciences.

Ralph W. F. Hardy is the president of the Boyce Thompson Institute, Cornell University, Ithaca, New York, and deputy chairman, BioTechnica International Inc., a biotechnology firm. His research interests focus on plant biology, biochemistry of nitrogen fixation and carbon input into crop plants, and photosynthesis. BioTechnica is active in developing products of biotechnology. He has served on numerous committees of the NAS/NRC and other national and international panels.

Edward L. Korwek received his doctorate in biochemistry from the University of Pittsburgh and his law degree from Duquesne University Law School. He is a partner with the law firm of Hogan & Hartson in Washington, D.C., where he specializes in regulatory law. He has served on the NIH Recombinant Advisory Committee and is currently on the Agriculture Biotechnology Research Advisory Committee. He is a on the Editorial Advisory Board of Biotechnology Law Report.

Richard E. Lenski (Chairman, Subcommittee on Microorganisms) is an associate professor of ecology and evolutionary biology at the University of California, Irvine. Dr. Lenski has conducted studies in population and evolutionary genetics, including applications that relate to stability and the fate of genetically engineered microorganisms. He was a consultant to the Office of Technology Assessment and was also one of the authors of the April 1989 Ecological Society of America document on environmental release of genetically engineered microorganisms.

Eugene W. Nester is a professor and chairman of the Deparment of Microbiology at the University of Washington, Seattle. His research interests include bacterial-plant interactions and the genetic engineering of higher plants by *Agrobacterium*.

Stanley J. Peloquin (Chairman, Subcommittee on Plants) is the Campbell Bascom Professor of Horticulture and Genetics, University of Wisconsin, Madison. Dr. Peloquin's research is on cytogenetics and evolution, and he is a specialist in potato breeding. He is a member of the National Academy of Sciences.

Calvin O. Qualset is a professor in the Department of Agronomy, University of California, Davis, and director of the California Genetic Resource Conservation Program. His research is directed at the genetics and evolution of disease resistance in plants, analysis of quantitative genetic variation in plants, breeding of barley, wheat, oats, triticale and rye, and genetic resources conservation in plants. He has served on numerous NAS/NRC committees and has been the president of the Crop Sciences Society of America and editor in chief of its journal.

Ralph S. Wolfe is a professor in the Department of Microbiology, University of Illinois, Urbana. He has studied the metabolism and physiology of bacteria including methanogens and archaebacteria. He is a member of the National Academy of Sciences.

SUBCOMMITTEE ON MICROORGANISMS

Richard E. Lenski (listed with the Steering Committee), Chairman

Peter J. Bottomley is a professor of microbiology and soil science at Oregon State University, Corvallis. Dr. Bottomley has conducted studies in the physiology, biochemistry, and ecology of nitrogen-fixing microorganisms in both laboratory and field situations. He has been a consultant to USDA-AID and to the Plant Division of the Oregon Department of Agriculture. He is a member of the editorial board of Applied and Environmental Microbiology.

Ananda M. Chakrabarty is a professor of microbiology at the University of Illinois College of Medicine, Chicago. Dr. Chakrabarty has conducted studies in biomedical sciences, in molecular cloning and genetic engineering with plasmids, and in the genetic basis of hydrocarbon biodegradation. He has served on committees for the NAS/NRC, the Society of Industrial Microbiology, COGENE, the European Economic Community, and the European Molecular Biology Organization. He is a Chairman of the Panel of Scientific Advisors of the United Nations Industrial Development Organization (UNIDO) International Center of Genetic Engineering and Biotechnology.

Rita R. Colwell is the director of the Maryland Biotechnology Institute, director of the Center of Marine Biotechnology, and professor of microbiology at the University of Maryland, College Park. Dr. Colwell has conducted studies in marine biotechnology, marine and estuarine microbial ecology, survival and ecology of pathogens in the marine environment, deep-sea marine microbiology, microbial degradations, and release of genetically engineered microorganisms into the environment. She has served as vice-president for academic affairs at the University of Maryland, past president for the American Society for Microbiology, and on numerous NAS/NRC committees. Dr. Colwell is president-elect of Sigma Xi, a member of the National Science Board, and vice president of the International Union of Microbiological Societics.

studies in plant pathology as related the physiology, genetics, and molecular biology of *Agrobacterium* spp. As part of these studies Dr. Farrand has been using genetic technology to engineer improved biological control systems for crown gall. He has served on U.S. Department of Agriculture and National Institutes of Health review panels, NRC workshops on plant-microbe interactions, and professional society committees.

Robert Haselkorn is the F.L. Pritzker Distinguished Service Professor in the Departments of Molecular Genetics and Cell Biology, Biochemistry and Molecular Biology, and Chemistry at the University of Chicago. He is also the director of the Center for Photochemistry and Photobiology at the university. Dr. Haselkorn has conducted studies of nitrogen fixation in cyanobacteria and in photosynthetic bacteria, and of the function of nucleic acids in viruses and cellular organelles, and of virus structure. He has served on editorial boards and in the virology study section of the National Institutes of Health. He is the past president of the International Society for Plant Molecular Biology and a member of the Panel of Scientific Advisors for UNIDO's International Center for Genetic Engineering and Biotechnology.

Roger D. Milkman is a professor of biology at the University of Iowa, Iowa City. Dr. Milkman has conducted studies in nucleotide sequence polymorphism, selection theory, genetic structure of species, population genetics, evolution, and molecular evolution. He has served with the genetics study section of NIH and on numerous editorial boards. He has been the secretary of the American Society of Naturalists and the secretary of the Society of General Physiologists.

Luis Sequeira is the J.C. Walker Professor in the Department of Bacteriology and Plant Pathology, University of Wisconsin, Madison. Dr. Sequeira has conducted studies in the molecular biology associated with virulence and avirulence in plant-pathogenic bacteria, host-parasite interactions in bacterial diseases of plants, plant growth regulators, and the physiology of parasitism. He is a member of the National Academy of Sciences. He has served on national and international committees and on numerous editorial boards; he was manager of the USDA Competitive Grants Office and is past president for the American Phytopathological Society.

James M. Tiedje is a professor of microbial ecology at Michigan State University, East Lansing. Dr. Tiedje has conducted studies in microbial ecology including denitrification, microbial metabolism of organic pollutants, and molecular methods to study soil populations. He has served on NSF, USDA and EPA panels and was editor-in-chief of *Applied and Environmental Microbiology.* He is chairman of EPA's Science Advisory Panel that has reviewed all proposed U.S. field tests of genetically modified microbial pesticides. He was an author of the April 1989 Ecological Society of America document on environmental release.

SUBCOMMITTEE ON PLANTS

Stanley J. Peloquin (listed with the Steering Committee), Chairman

Roger N. Beachy is a professor of biology at Washington University, St. Louis, Missouri. His research focuses on the control of synthesis of soybean seed proteins, plant viral messenger RNAs, effects of virus gene products on infected host cells, and the genetic transformation of plants for virus resistance. He is a fellow of the American Association for the Advancement of Science.

Donald N. Duvick is vice president for research at Pioneer Hi-Bred International, Inc., Johnston, Iowa. His research interests are in cytoplasmic inheritance of pollen sterility in maize, immunological identification of plant proteins, developmental morphology and anatomy, and genetics. Dr. Duvick is past president of the National Council of Commercial Plant Breeders and of the Crop Science Society of America.

Robert T. Fraley is manager of plant molecular biology at the Monsanto Company, St. Louis, Missouri. In addition, he is an adjunct professor at Washington University, St. Louis. His research interests are in genetic engineering of plants and the development of efficient systems for introducing and monitoring the expression of foreign genes in plant cells. He has had editorial responsibility for several professional journals.

Richard N. Mack is a professor and chairman of the Department of Botany, Washington State University, Pullman. His research interests are primarily in the invasions of vascular plants in North America, plant population biology, and community ecology. Dr. Mack is a member of the Ecological Society of America.

Anne Vidaver is a professor and head of the Department of Plant

Pathology, University of Nebraska, Lincoln. Her research interests are in phytopathogenic and beneficial bacteria, bacteriophages, and bacteriocins. Dr. Vidaver is an alternate member of the USDA's Agricultural Biotechnology Research Advisory Committee. She is a fellow and has been president of the American Phytopathological Society; she is a fellow of the American Association for the Advancement of Science.